## Pre-title sequence

It's dark. Everything is calm and quiet except for the lapping of waves against the shore, the distant chink of champagne glasses and occasional snatches of laughter caught on the breeze. The agent emerges from the water and wades up the beach. He glances round before peeling off his specially made rubber over-suit to reveal the tuxedo he is wearing underneath. He spots his contact, who rushes over to whisper some hurried words and sprinkle him with a few drops of Hennessy brandy, the final touch needed to give the impression of a drunken party-goer. As prepared as he ever will be, the agent strides towards the seafront casino and introduces himself.

Audiences have certain expectations when they sit down to watch or read a James Bond adventure. Secret missions, daring stunts, big explosions and evil plots may be part of any spy, thriller or action film. But there is something unique about Bond, something the filmmakers themselves came to term 'Bondian'. There is no other franchise like it. A secret agent risking his life swimming through hostile waters to infiltrate enemy territory dressed in a dinner jacket and black tie is exactly what you would expect of an opening sequence in a 007 film. It is how one of the most iconic Bond films, *Goldfinger*, introduces itself to audiences. It is also real life.

The man with the evening suit making his stylish way towards the casino in the opening paragraph was Tazelaar,

Pieter Tazelaar. He was a member of the Dutch Resistance who was dropped offshore at Scheveningen in Nazi-occupied Netherlands at 0435 hours on 23 November 1940. The mission, the boat that dropped him there, the rubber suit that kept his tuxedo dry, had all been arranged by the British secret service.

The world of James Bond is a fantasy of larger-than-life villains, plot holes you could steer Stromberg's supertanker through and cars overstuffed with so many gadgets they should be undriveable. But, without some grounding in real life, some grains of truth at its heart, or acknowledgement of the real world, it would become an absurd parody of itself.

The well-worn formula of girls, quips, guns and plans to take over the world has attracted an estimated quarter to a half of the world's population to watch a Bond film, and over 100 million people to read the books. Part of the joy of watching and reading these stories is living a thrilling vicarious life from the comfort of our sofa, basking in their glamour but untroubled by the danger from which our fictional hero is always fighting to extricate himself.

There will be times in this book when I write about Bond and his world as though they are real. This will annoy some but be completely understandable to others. Many of us have fantasised about living Bond's exciting life or speculated what we might have done in his position. But with 007's adventures deliberately paced at breakneck speed, it is only after putting down the book or staggering out of the cinema at the end of the film that we can catch our breath and start to think about what just happened.

This book is an extended version of those conversations and speculations after watching a Bond film, told from the point of view of a fan and a scientist. Film by film and iconic moment by iconic moment, this book explores some of the many tropes that we have come to love and expect from a Bond adventure. So, with cuffs straightened, vodka martini in hand and tongue firmly in cheek, let's take aim straight down the gun barrel and head off on our mission to explore the science of James Bond.

# SUPERSPY SCIENCE

## Science, Death and Tech
## in the World of James Bond

Kathryn Harkup

BLOOMSBURY SIGMA
LONDON · OXFORD · NEW YORK · NEW DELHI · SYDNEY

*To Bill Backhouse*

BLOOMSBURY SIGMA
Bloomsbury Publishing Plc
50 Bedford Square, London, WC1B 3DP, UK
29 Earlsfort Terrace, Dublin 2, Ireland

BLOOMSBURY, BLOOMSBURY SIGMA and the Bloomsbury Sigma logo are
trademarks of Bloomsbury Publishing Plc

First published in the United Kingdom in 2022

A catalogue record for this book is available from the British Library

Library of Congress Cataloguing-in-Publication data has been applied for

ISBN: HB: 978-1-4729-8226-1; TPB: 978-1-4729-8227-8; eBook: 978-1-4729-8223-0

2 4 6 8 10 9 7 5 3 1

Typeset by Deanta Global Publishing Services, Chennai, India

Printed and bound in Great Britain by CPI Group (UK) Ltd, Croydon CR0 4YY

Bloomsbury Sigma, Book Seventy Five

To find out more about our authors and books visit www.bloomsbury.com
and sign up for our newsletters

# Contents

Pre-title sequence   1

001: *Dr No* and the gun-barrel sequence   9

002: *From Russia with Love* and Rosa Klebb's shoe   24

003: *Goldfinger* and the laser   40

004: *Thunderball* and the gamma gas   52

005: *You Only Live Twice* and the volcano lair   68

006: *On Her Majesty's Secret Service* and Blofeld's
bioterrorism plot   83

007: *Diamonds Are Forever* and diamonds   97

008: *Live and Let Die* and the crocodile run   111

009: *The Man with the Golden Gun* and the
golden gun   127

010: *The Spy Who Loved Me* and the
parachute jump   143

011: *Moonraker* and the exploding space station   157

012: *For Your Eyes Only* and electrocution
through headphones   171

013: *Octopussy* and the atomic bomb   186

014: *A View to a Kill* and May Day   201

015: *The Living Daylights* and the cello case 215

016: *Licence to Kill* and a tanker full of cocaine 230

017: *GoldenEye* and the EM pulse 246

018: *Tomorrow Never Dies* and the stealth boat 260

019: *The World Is Not Enough* and Renard's bullet 274

020: *Die Another Day* and being sucked
out of a plane 291

021: *Casino Royale* and the knotted rope 304

022: *Quantum of Solace* and the girl covered in oil 320

023: *Skyfall* and the cyanide capsule 335

024: *Spectre* and Bond's backstory 348

025: *No Time to Die* and the nanobots 363

Bibliography 379

Acknowledgements 387

Index 389

# 001

# *Dr No* and the gun-barrel sequence

A white dot moves across the screen. A man appears at its centre. The circle follows him as he walks. Suddenly, he turns, aims a gun and shoots. Red washes down the screen and loud, familiar notes blast out from the speakers. Bright colours, abstract shapes and silhouettes of writhing women appear. Within seconds we know exactly where we are and what to expect. More or less.

The James Bond theme, the gun-barrel sequence, the highly stylised titles – it was all there from the start. The only things missing from the first Bond film were a pre-title sequence and a title song. The following 24 films would be introduced in almost the same way, rearranged and reinterpreted, but instantly recognisable as part of the same franchise. Each new cinematic addition is simultaneously comfortingly familiar and new and exciting. It is an impressive trick the producers pull off: repeatedly reinventing a highly formulaic series of films that draws audiences of millions back for more, time and time again.

*Dr No* started what would become a familiar pattern over the following 60 years. The film opens with an intrigue: the assassination of a man and woman who both work for the secret service. Bond is briefed by M and equipped for his mission by the Quartermaster (see chapter 002): in this film he is given his trademark gun, the Walther PPK, and deprived of his favourite Beretta (see chapter 009). On his way out of the office he flirts with Moneypenny, then heads to an exotic location: in this case, Jamaica. He meets his contacts: Quarrel and Felix Leiter. Like a detective in a whodunit, he discovers clues everyone else has overlooked that hint at a plot much larger and more sinister than anyone had hitherto suspected: radioactive rocks from the mysterious Crab Key. Bond seduces Miss Taro, a beautiful woman working for the bad guys – a character type who, in future films, if she isn't converted to the good side by Bond's sexual charms, will almost certainly be killed and often in an unusual way. Bond meets Honey Ryder, another beautiful woman- but she is on the side of good. She becomes instantly loyal to him but won't succumb to his advances until the closing credits (see chapter 014). Along the way there are several half-hearted attempts on Bond's life, but he must be kept alive and in sufficiently good shape to meet the main, heavily stereotyped villain (see chapter 019): Dr Julius No, a power-crazed Chinese-German with mechanical hands. The bad guy is living in very comfortable and technologically well-appointed seclusion (see chapter 004): a subterranean palace located on Crab Key island with a dragon (see chapter 018) and swamp to deter visitors, but also luxurious guest accommodation and prison cells for those who can't take the hint, along with a nuclear power plant on the side (see chapter 012). With elaborate courtesy, the villain explains his plan for world domination (see chapters 006, 008): disrupting rocket launches from the USA using some kind of nuclear-powered radio transmitter. And the villain does all this before trying to kill Bond in an extravagant and convoluted way that enables his escape (see chapters 003, 021). The film finishes with a spectacular set piece involving lots of explosions (see chapter 011) and Bond getting the girl. Bond escapes Dr No's lair as it blows up, thanks to his interference with the nuclear reactor, grabbing Ryder along the way, and they sail off together in a little boat. They finally kiss as the credits roll.

There is still time for cocktails, sardonic quips, daring stunts, exciting chases, magnificent sets and outrageous sexual innuendo, and all in a very entertaining, fun, fast-paced two(ish)-hour film. There is, of course, the secret agent at the centre of it all, James Bond, but he is the focus of later chapters (021 and 024). There are many other not so obvious factors that go into making a Bond film characteristically 'Bondian'. One of the most notable features of any 007 film is the supreme self-assurance that oozes from every frame. Bond and everyone involved know exactly what they are doing.

Such confidence came from the producers' firm belief in the project and certainty as to how best to present this character and his world on the big screen. It was also, in no small part, thanks to the existence of several novels and their author Ian Fleming being on hand to offer advice, background and ideas for what 007 was and should be. James Bond and his world are the products of elements from Fleming's own life, his fertile imagination and lashings of exaggeration.

### Setting the scene

Ian Lancaster Fleming was born on 28 May 1908 to Valentine (Val), a Conservative MP who died a war hero in 1917, and Evelyn, who dominated Ian's life, especially after the death of her husband. Fleming attended Eton and Sandhurst but left both under a cloud.[1] His mother sent him to a finishing school in Kitzbühel, Austria, run by Ernan Dennis-Forbes, a former Secret Intelligence Service (SIS) employee, and his wife, the writer Phyllis Bottome. It was here that Fleming learned to ski and, with Bottome's support, wrote what he considered to be his first story, 'A Poor Man Escapes'. The plan had been for Fleming to enter the Foreign Office but he failed the entrance exams. Instead, his mother got him a job with Reuters, where he was able to hone his writing skills, travel and get caught up in exciting international intrigues such as the Metropolitan-Vickers trial in Moscow.[2] Ian seemed to have found a role that suited him but, bowing to family pressure, he gave it up. He went into banking then stockbroking, jobs he obtained based on his family owning a bank rather than any specific talent or qualifications, and was terrible at both.

Salvation for Fleming came in the form of the Second World War. He was recruited to work as personal assistant to the head of the Naval Intelligence Division (NID), Admiral Godfrey. His charm and connections made him ideal for the role although he had little in the way of actual qualifications or experience.[3] He liaised with different branches of the British Intelligence services – MI5, SIS (what became MI6) and the Special Operations

---

[1] Eton because of his over-enthusiasm for hair oil, cars and women. Sandhurst because he contracted gonorrhoea.

[2] Six employees of the British firm were accused of spying, but more of this in chapter 002.

[3] Though speaking French, German and a smattering of Russian were a definite asset.

Executive (SOE)[4] – and was privy to detailed information on secret intelligence and special operations. He was trusted enough within the organisation that he could make suggestions, some sensible that were taken on board and others that were outlandish and better suited to the novels he would write a decade later.[5]

Though Fleming was at the heart of many intelligence operations during the war, and met many people who would provide inspiration for his daring fictional spy hero, he was never personally involved in a mission. The risk of his being captured and divulging secret information to the enemy was too great. His high alcohol and cigarette consumption also ruled him out on health reasons, traits he passed on to his fictional spy, minus the debilitating hacking cough or hangovers.

At the end of the war Fleming returned to a comfortable civilian life. He became manager of the foreign correspondents at the Keylsey newspaper group,[6] which included the *Sunday Times*, where he contributed columns. But while the rest of the UK was enduring the worst of the British weather, Fleming negotiated three months' annual leave every winter, which he spent at his second home in Jamaica. He had told several people during the war that he was going to 'write the spy novel

---

[4] SOE was set up during the war to carry out clandestine operations in Europe, such as parachuting people and equipment into occupied territory.

[5] Contrast recruiting a team of commandos (30 Assault Unit or 30AU) to retrieve important German documents, with the proposal to recruit Aleister Crowley for his expertise in black magic.

[6] Foreign correspondents were sometimes used as cover for MI6 work at this time.

to end all spy novels' and, on the 'third Tuesday in January' 1952, Fleming sat down at his typewriter and started *Casino Royale* – 'to take my mind off the shock of getting married at the age of 43'. He finished the first draft in two months and the book was published the following year. He soon settled into a routine of writing every morning during his Jamaican stay, churning out 2,000 words a day and a novel every year.

In Jamaica he let his imagination run riot. It was when he returned to the UK that he would smooth out the rough draft and fill in the details. To tether his outrageous plots to some semblance of reality, he included lengthy and accurate descriptions of cars, food and branded consumer goods as well as credible nuggets of information about the secret service and spy craft, while being careful not to reveal any genuine secret information (for example, Bond's well-known cover story of working for Universal Exports is based in fact). From its establishment in 1909, the British secret service took the 'secret' part of its name very seriously. All sorts of elaborate cover stories and subterfuge were constructed to maintain the illusion of the service's non-existence. At the very start a 'cover address' was set up for correspondence: 'Messrs. Rasen, Falcon Ltd., Box 400, General Post Office, London' – a supposed shipping and export firm that set the trend for 'import and export' as a cover for espionage work.

To add further realistic flesh to the bones of his fantasy, Fleming also borrowed real-life locations. A health clinic he visited ended up in *Thunderball*, and his golf club, renamed, features heavily in *Goldfinger*. The setting of Dr No's Crab Key lair was inspired by a trip to Inagua, a 'hideous island' that 'nobody in his senses ever goes near', in the southernmost reaches of the Bahamas. It had last been officially surveyed in 1916, since when it had disappeared off the bureaucratic radar. Fleming and his friend Ivar Bryce were invited to join a scientific expedition studying the island's huge population of flamingos.[7] It wasn't the enormous numbers of pink birds that Fleming remembered most vividly, but the 260km$^2$ (100mi$^2$) shallow lake, 'the colour of a corpse' and stinking of rotten eggs, that they lived on. The flamingos shared the island with around a thousand people, almost all of them employed in the island's only industry, a saltworks owned by three brothers who ruled over the island as their own private kingdom. It was the perfect location for a Bond villain to set up his criminal operation.

Characters and names were appropriated from Fleming's vast network of friends and acquaintances. M was based on his former NID boss, Admiral Godfrey, who was known for his long face and look of permanent displeasure. The moniker 'M' was probably taken from the head of MI6 being referred to by a single letter, a tradition started by its original head Mansfield Cumming, who signed himself 'C'.[8] Fleming simply switched the letter to M, perhaps after Sir Stewart Menzies, head of MI6 during the Second World War, or maybe after his mother, who signed her letters 'M'.

---

[7] Fleming and his party travelled around the island on a marsh buggy that would become the fire-breathing dragon tank of the novel.

[8] Every head of MI6 since has been known as 'C'.

Some people may have been flattered by the characters chosen to sport their names. Hilary Bray, a golfing buddy, became a heraldry expert in *On Her Majesty's Secret Service*. Ernie Cuneo, an American friend, appeared as the ill-fated taxi driver in *Diamonds Are Forever*. Others may not have seen it as a compliment. Tom Blofeld[9] was at Eton with Fleming and became immortalised as Bond's nemesis Ernst Stavro Blofeld. Lord Arran, a cousin of Fleming's wife Ann, known to friends as 'Boofy' Gore, appeared in *Diamonds Are Forever* as a murderous hood. Lord Arran objected so strongly that an apology was given and the name changed in all subsequent editions of the book. The architect Erno Goldfinger consulted his lawyers after his name was appropriated for a Bond villain. On this occasion, Fleming refused to back down. He responded: 'Tell him that if there's any more nonsense I'll put an erratum slip and change the name throughout to GOLDPRICK and give the reason why.'

The stories Fleming concocted on his golden typewriter[10] were, he admitted, 'the feverish dream of the author of what he might have been – bang, bang, kiss, kiss – that sort of stuff.' It was a world that appealed to a lot of other people too, and his blend of realistic detail and overblown fantasy struck a chord with tens of thousands of readers. Sales were helped enormously by Fleming making excellent use of his extensive contacts to secure reviews in high-profile papers and magazines.

---

[9] Father of the cricket commentator Henry 'Blowers' Blofeld.

[10] He really did write on a specially commissioned gold-plated typewriter. Many of his friends thought it a vulgar affectation. Fleming didn't care.

But, while sales were healthy, the books were not exactly a publishing phenomenon. All that changed after the release of the first Bond film, *Dr No*.

Fleming had been involved in many discussions with filmmakers and TV executives over the years. *Casino Royale* had been adapted for a one-hour TV special in 1954 with Barry Nelson as an American Bond. Fleming had also been commissioned to plot out episodes for other spy-themed TV series, and options had been taken on several of the Bond books. But nothing else had made it into production, until Albert 'Cubby' Broccoli and Harry Saltzman appeared on the scene. In 1961, Saltzman had the rights to all the Bond novels except *Casino Royale*, but no backers.[11] Broccoli desperately wanted to make a Bond film but didn't have the rights. Fortunately the two met and the rest, as they say, is history.

## Establishing the rules

Initially, the producers were spoiled for choice. They had eight books at their disposal, all packed with ideal fodder for films: sex, violence, action and intrigue. They chose *Dr No*, the fifth novel, published in 1958, for a few reasons. Their preferred choice, *Thunderball*, was caught up in lengthy legal disputes.[12] *Dr No* was based on an

---

[11] Saltzman's daughter suggested Harry may have got the film rights because of a mutual understanding between Fleming and himself; they had both been involved in secret missions during the war. Saltzman had done classified work for the Psychological Warfare Division of the US Office of War Information. Some suspect the two had first met much earlier than 1961.

[12] *Thunderball* started life as a film treatment co-written with Kevin McClory, Jack Whittingham, Ivar Bryce and Ernst Cuneo. When it

idea for a TV series that never got made and so much of it had been written with a view to how it would appear on screen. And there were plenty of other aspects that were a natural fit for film.

Exotic locations always feature prominently in the James Bond books. Fleming loved to travel and presumably readers loved to read about far-flung places at a time when foreign holidays were an unobtainable luxury for most. The sun-drenched, sandy shores of Jamaica must have looked particularly appealing to audiences watching *Dr No* when it was released in the UK in October 1962.

*Dr No*'s titular character, like so many of Fleming's villains, has more than enough ego to fill the big screen.[13] His plan, to divert rockets launching from Cape Canaveral, is perfect for the 1960s, when the space race was in full swing. The radio jamming/guidance system he uses got a nuclear upgrade in keeping with atomic preoccupations of the age. It was an obvious path to take at a time when world leaders talked about technological advances with enthusiasm and optimism. John F. Kennedy spoke of the 'new frontier' as the United States set their technological sights on space and the moon. Harold Wilson in the UK talked about 'the white heat of science and technology'. As the Bond series progressed, science

---

failed to go into production, Fleming used many of the ideas for his next novel. McClory and Whittingham sued. Their names were added to subsequent editions of the novel and they gained the rights to be involved in any film production should it come about.
[13] Fleming suggested his friend Noël Coward for the role of Dr No, to which Coward responded by telegram, 'Dr No? No! No! No! No!'

and technology became a fundamental part of the movies. The filmmakers have continuously adapted and updated the science to keep pace with rapid developments and emerging areas of research in the real world. They have kept up with, and often exceeded, the technological capabilities of the day, to entice audiences into the cinema to witness up-to-the-minute gadgets and the latest cars.

Topicality is important to the films, but it also dates them. A joke about a stolen Goya painting in *Dr No* is lost on modern audiences but apparently brought the house down in 1962.[14] References to solar power, as a way out of the energy crisis that was happening at the time *The Man with the Golden Gun* was released, still resonate today, but for different reasons. Though these were deliberate additions, sometimes real life has overtaken the imaginary world of Bond. *Licence to Kill*, a film about a South American drug cartel, hit cinema screens six months before Manuel Noriega was arrested (see chapter 016). The release of *No Time to Die*, a film about a deadly infectious agent spreading across the world, was delayed because it coincided with an outbreak of a deadly infectious virus spreading across the world. Audiences first watched Dr No attempt to send American rockets off course a few weeks before the Cuban Missile Crisis.

Though the films are undoubtedly products of the post-Second World War and Cold War era, and later a

---

[14] The painting was stolen from the National Gallery in London on 21 August 1961 and wasn't recovered until July 1965. But it was a retired bus driver Kempton Bunton who did the deed, not Dr No.

post–Cold War world, the producers were keen to avoid politics. In the novel, Dr No is working for the Russians. Fortunately for the filmmakers, the novels contained a ready-made apolitical criminal organisation that was big enough and bad enough to take on any evil plan. SPECTRE (Special Executive for Counter-Intelligence, Terrorism, Revenge and Extortion), a kind of international private members' club for megalomaniacs, was, I'm sure, more than happy to welcome Dr No into its villainous ranks. He already had the distinctive lair, exotic pet and unhealthy obsession with power and money that form the basic entry requirements. To keep audiences happy, and the films acceptable to almost any political persuasion, frightening or mundane reality is swept under a fantasy carpet or reshaped into a more appealing form.

Britain in the 1950s and 60s was still economically and physically scarred by the Second World War. Its once-vast empire was fracturing and breaking away from British rule. The Suez crisis of 1956 had been a political embarrassment. The defection of Guy Burgess and Donald Maclean in 1951, and Kim Philby in 1963, left doubts over the effectiveness and security of the British secret service. But Fleming's James Bond exists in a world where the British empire hasn't crumbled, the UK is still a major influence in global politics and individual, plucky British heroism can save the day.

The novels contain all the raw material the filmmakers needed to build the escapist fantasy world of 007 we see in cinemas, but they can veer wildly between the graphically realistic and the inexplicably daft. More than a little refining was needed for the big screen. The screenplay for Dr No went through many iterations before the producers were satisfied. The science and technology got a more prominent role and some of Fleming's worst excesses were trimmed back. Bond's big fight with a giant squid was cut and Dr No's death under a pile of bird poo was changed to drowning in the coolant of his nuclear reactor.[15] Even a franchise that would later show Margaret Thatcher, Prime Minister of the United Kingdom, in conversation with a parrot, had its limits.[16] Scenes were cut, new ideas were pasted in and some things simply got lost in translation. In the film, Bond escapes from Dr No's prison cell through the ventilation system, which has gallons of water sloshing around in it for no obvious reason other than it's exciting to watch Bond avoid being washed away in the flood. In fact, it was a hangover from the book, in which Dr No has deliberately installed an elaborately booby-trapped maze of tunnels out of his prison cells so he can watch his captives suffer as they try to escape.

The aim was to make the film as exciting and entertaining as possible without confusing the audience, but the result doesn't always make sense. Why, in the gun-barrel sequence, are we looking down the barrel and not through the sights? How does so much blood get inside the gun when gunshot victims in the rest of the film barely bleed? How can someone clever enough to build a nuclear reactor on a remote swamp-infested

---

[15] Or is he boiled to death? Or frozen? Maybe he's electrocuted or killed by the radiation. From what we see on screen, I honestly can't tell, but it looks unpleasant.

[16] An early draft, where Dr No is revealed to be not the tall, skinny, yellow-skinned man of the book, but a spider monkey sitting on his shoulder, was rejected immediately.

island have such poor safety controls that his whole
operation can be sabotaged by Bond turning a wheel?
The 007 filmmakers had to use a number of tricks to
gloss over the improbabilities.

First of all, the colours are brighter and the sound
effects louder to help audiences understand this is an
exaggerated comic-book world. The fast editing doesn't
give anyone time to think too deeply. If the transitions
aren't too jarring or the ideas too extreme, no one notices.
And, as long as it's still entertaining, no one much cares
either. Even so, it takes a lot of skill to sweep an audience
along for the ride, and it is easy to get it wrong.[17] Judging
by the success of *Dr No*, and all the Bond films since,
Broccoli, Saltzman and the team that continue to make
them consistently get the balance right.[18]

For films that are backed by American dollars,
produced by an American and a Canadian[19] and feature
a notably international cast, they are very British. It is
fascinating and bewildering that a film series so steeped
in British patriotism, and often unflattering to other
nations, remains so popular around the world. It is just
another of the many contradictions, sharp contrasts and
oddities that the 007 franchise appears to thrive on.

**Bond will return**
The now-trademark teaser line shown as the credits roll
didn't appear in *Dr No*, though the film's massive success

---

[17] See the 1967 adaptation of *Casino Royale*.

[18] Of course, everyone's mileage will differ.

[19] And all American producers since Saltzman's departure from the
franchise.

ensured that he, and many other, would try and emulate 007 and his adventures. As the sixties progressed, many tried to jump on the 'Bondwagon', as it became known, with serious and not-so-serious imitations and interpretations of the spy adventure. In 1965 there were more than 50 spy-themed films released. The following year there were more than 60. But 007's influence stretched beyond stories of espionage. In the past six decades there is scarcely a thriller, action film or adventure story that doesn't owe something to the big-screen James Bond, be it a suave hero, daring stunts or quick editing.[20]

The film version of *Dr No* is largely faithful to the book, but more changes had to be made as time went on and the world changed. The films have drifted further from Fleming's original as the source material – just 14 books – has been exhausted. But even after 25 films, there is still a direct link to the original elements within Fleming's stories and the style and standards set by *Dr No* – big explosions, exciting chases, fast cars, beautiful women and villains threatening the world.

---

[20] It can also be sombre-toned, gritty realism, such as *The Spy Who Came in From the Cold*, written as the antithesis of James Bond.

# 002

# *From Russia with Love* and **Rosa Klebb's shoe**

Where would 007 be without his gadgets? Still able to save the day, obviously, as *Dr No, On Her Majesty's Secret Service* and *Quantum of Solace* proved. But, when your hero is trapped behind bars, having a fountain pen full of acid does help move the plot along (*Octopussy*). Sure, there are lots of ways of creating a diversion, but an exploding dart fired from a cigarette will certainly take everyone by surprise (*You Only Live Twice*). Gadgets have both helped and hindered Bond in his adventures, and not just because these ingenious little devices are often in the hands of his enemies. His fellow spies are equally well equipped whether they work for the CIA (Goodhead in *Moonraker*), KGB (Amasova in *The Spy Who Loved Me*) or Chinese secret service (Lin in *Tomorrow Never Dies*). The possession of a gadget instantly associates a character with the world of espionage rather than an innocent bystander simply caught up in events.

*From Russia with Love*[1] was the first of the Bond films to feature gadgets. It starts with head henchman, Red Grant, stalking through a hedge-maze with a garrote wire concealed in his watch. If this is what Bond is up against he is going to need at least a cannister of tear gas hidden in a tin of talcum powder to level the playing

---

[1] The film title drops the comma from Fleming's novel, *From Russia, with Love*.

field. But he is also given 50 gold sovereigns (for bribes), a throwing knife, a rifle with infra-red sights that can be packed into its own stock, 20 rounds of ammunition and a film camera with a hidden voice recorder. And all this is squeezed into one briefcase, expertly prepared by Q, who also makes his first appearance.[2] The baddies aren't relying on a single garrote wire either. They have another weapon that sticks in the audience's mind, and the shins of its victims – the poison-tipped knife hidden in Rosa Klebb's shoe. On top of all that, the entire plot revolves around obtaining another gadget – a Lektor coding machine, used by the Russians to encrypt their most secret messages.[3]

The following film, *Goldfinger*, has so many gadgets they no longer fit inside a briefcase and must be packed into a car, the classic Aston Martin.[4] From then on, gadgets became a much-anticipated part of every film. The grown-ups get to enjoy speculating as to how these gadgets will be used, and younger audiences can add the smaller, safer, toy versions to their Christmas lists.

What constitutes a gadget is very much open to debate, and not just in the 007 films. Are they different to devices and gizmos? And what about widgets and thingummybobs? Everyone you ask will have a slightly different answer but, generally, they are expected to have a specific but limited use, often ending up forgotten in a

---

[2] An armourer, Major Boothroyd featured in *Dr No*, but this is the first time the character is referred to as Q.

[3] Called a 'Spektor' in the novel but renamed for the film to avoid confusion with a certain international crime syndicate.

[4] There are far too many car- and vehicle-based gadgets to fit in here, so they will be covered in chapter 018.

drawer (see Bond's desk in *On Her Majesty's Secret Service*, or Q's workshop in *Die Another Day*, littered with relics of gadgets from previous films that have become obsolete). A gadget may be one device disguised as another (see the perfume/flamethrower in *Moonraker*). It may appear useful but it isn't always (Bond's electromagnetic watch in *Live and Let Die* that fails to rescue him from a pond full of crocodiles). There is also an element of absurdity about a gadget (see the flame-throwing bagpipes in *The World Is Not Enough* or the man-eating, revolving sofa in *The Living Daylights*). Does size matter? Can George Lazenby's combination safe-cracker and photocopier be dismissed because it is so big it must be winched through a window by crane? This chapter probably doesn't have the answers, but it will look at some of those iconic items that have featured so memorably in many of the films and books.

## The right equipment

In a preface to *From Russia, with Love*, Fleming made a point of tying events in the book to real life by asserting his knowledge of Russia and the workings of its intelligence services. In 1933 he had been sent by his employers, Reuters news agency, to cover the Metropolitan-Vickers trial in Moscow. One of the company's secretaries, Anna Kutsova,[5] had been bundled into a car one morning on her way to work. Six British employees were also arrested on charges of espionage and spying and faced possible execution. In his

---

[5] Supposedly the inspiration for Fleming's fictional Tatiana Romanova in *From Russia, With Love*.

determination to get the scoop on all the other journalists covering the story, Fleming stooped to a few dirty tricks. He got chummy with a clerk in the censor's office and persuaded him to approve two versions of his article – one where the men were found guilty and one with a not-guilty verdict. He then paid a boy to stand under the window of the court room so he could drop the correct version to him and run it to the telegraph office.[6]

Fleming returned to Russia in 1939 to write a special intelligence report for the Foreign Office. He was seconded to *The Times* as cover and believed himself to be completely inconspicuous, or at least credible as a journalist rather than a spy, but probably wasn't fooling anyone. On top of his personal experiences in Russia, he also borrowed from real people and events that he had read about.

In the book, Bond is lured to Istanbul to retrieve a beautiful, defecting Russian agent and a Russian coding machine. When they leave Istanbul on the Orient Express, an assassin follows them and tries to kill Bond. The attempt on Bond's life was inspired by American Naval attaché Captain Eugene Karpe, who in 1950 was either pushed or fell to his death from the famous train.[7]

His fictional Russian agent, Rosa Klebb, was inspired by the real-life Russian intelligence officer Colonel Rybkin.[8] Fleming made her a member of SMERSH (a

---

[6] In the end he was pipped to the post by an American reporter who happened to be on the phone to his office when the verdict was announced over the building's tannoy.

[7] It's not just Agatha Christie bumping people off on the Orient Express.

[8] Fleming wrote an article about Rybkin for the *Sunday Times*. He considered her to be one of the most powerful women in espionage.

contraction of *Smiert Spionam* meaning Death to Spies), a department that he claimed, 'exists and remains today the most secret department of the Soviet government.' And though there really was such a department within the KGB, it had been disbanded in 1946, long before the book was written. He gave an address for their headquarters and asserted that his descriptions of their conference room, as well as some of the individuals who met there, were accurate.[9]

Then, to equip the agents on both sides, he again drew on his personal experiences.[10] During the Second World War Fleming had cultivated friendships with inventors and proudly possessed some of their products. Throughout the war years he carried a fountain pen with him loaded with tear gas, which could also take a cyanide cartridge for 'really dangerous missions'. The British secret service had long made use of scientists and engineers to support their work, but the war gave impetus to scale up and branch out into all kinds of devices and gadgets.

## Q and Q-Branch
Mansfield Cumming, the first head of the secret service, has been described as having a boyish enthusiasm for

---

[9] The one major change from book to film was to attribute all evil plans and any dirty tricks to SPECTRE, and nothing to do with the Russian government. Nevertheless, the film was banned in Russia for a long time, though illegal copies still made it through the Iron Curtain.

[10] Bond's briefcase, unlike in the film, also contains a cyanide pill in the handle; the talcum powder is instead a can of shaving cream hiding a silencer.

gadgets, much like Fleming, and encouraged scientific and technological research. Initially the service relied on expertise from the armed forces, but in 1919, 10 years after its founding, the service took on a Dr S. Dawson 'as chemist' on a consultancy basis. Over time more people were employed to look at 'secret inks, bombs, etc.'

During the Second World War, huge effort was put into cryptography, communication devices, as well as means of smuggling information, explosives, weapons and much, much more. People were needed to design and develop all of this and one of those people was Sydney Cotton, an Australian inventor, aviator and photographer. In later life, he recalled a visit to Fleming's flat to brainstorm how technical innovation could help with intelligence gathering. Then there was Robert Churchill, a gunsmith who worked 'on gas pistols and God-knows-what', who was the source of Fleming's poisonous fountain pen. Another acquaintance was Charles Frazer-Smith, the man who kitted out British agents with all sorts of gadgets. He commissioned over 300 firms around London to make everything from shaving brushes with secret cavities to shoelaces that acted as saws. Frazer-Smith said that Fleming had been particularly fascinated by his hollowed-out golf balls used to conceal messages to prisoners of war.[11] And there were many others who could also have provided inspiration for the character of Q and his specialist workshop.

When the war ended and a new, very different Cold War developed, the technological research was reassessed

---

[11] A trick Fleming used to smuggle diamonds in *Diamonds Are Forever*.

to meet the changing needs of the service. Two technical sections were established. One was charged with investigating matters such as explosives, weapons and various chemical tasks. The other focused on communications and electronic development. An experienced army quartermaster colonel, with the designation Q, was brought in to manage stores of equipment and transport. His team, Q-Branch, also took over photographic tasks, printing, reprographics and training, sections that had existed since before the war.

In September 1947, a series of newsletters were produced to inform colleagues about the type of work that was being carried out. Their challenges, 'to evolve items of equipment for specialised work', were not a million miles from something you might see in a Bond film. The first newsletter detailed several areas of interest including 'a device which will increase the security of operators on burglarous enterprises', which translates as a torch that produced a deep red light.[12] Another was 'a knockout ampoule or tablet which will behave in a reasonably predictable manner'. There was also interest in developing 'a method of opening combination safes'. They hoped to produce an electronic device rather than training their operatives with techniques like 'finding the gap' by sandpapering their fingertips.[13] Guns and silencers, however, were adapted to their specific needs rather than designed from scratch (see chapter 008).

---

[12] Infra-red equipment at the time was too heavy to be practical.

[13] As well as Lazenby's bulky safe-cracking machine, Connery uses a pocket-sized device to crack a safe in *You Only Live Twice*. And Pierce Brosnan got an even smaller safe-cracking gadget in *Tomorrow Never Dies*.

There was plenty of material for Fleming to work with and, combined with his enthusiasm for gadgets, it is surprising he didn't write about more of them. The novel *From Russia, with Love* is the exception rather than the rule. Cubby Broccoli was just as enthusiastic and, with his own connections to the military and an excellent props department supporting him, gadgets were not only included in the films but their number, technological abilities and importance to the plot was greatly increased.

**Pay attention, 007**

Though the film version of *From Russia with Love* doesn't show Q's workshop, all the other well-known elements are there. Desmond Llewelyn as Q battles to get Bond to concentrate as he explains important details about the equipment he and his team have carefully put together for 007 to use, and often destroy, later in the film.[14] Though Bond is usually dismissive of these inventions (and in this film he claims he probably won't need them), he always finds some use for them even if it is not quite what Q intended. The tear gas and knife loaded into the attaché case are both used to fight off the formidable assassin Red Grant, but Bond eventually kills him with Grant's own gadget – the garrote wire in his watch.

When Grant fails, Klebb is sent to kill Bond. She attacks him in his hotel room when he has none of his gadgets, or even his gun, available to help him. Klebb may be tiny but she is vicious. Bond must fend her off with a chair. But it's not enough to keep him out of

---

[14] Roger Moore regularly pranked Llewelyn, handing him pages of invented technical dialogue that the actor struggled to memorise.

range of Klebb's secret weapon, the blade in her shoe. What the audience knows, and Bond doesn't, is that the blade will deliver more than a nasty wound; its tip is covered in a venom that can kill in 12 seconds.

For thousands of years weapons have had their lethality enhanced by the addition of toxic substances. Poisons have been added to arrow heads and smeared on swords. As weapons technology advanced, deadly substances were not abandoned, but refined. The Nazis investigated adding cyanide to bullets, and pellets fired into Bulgarian dissidents have been filled with ricin. The knife part of Klebb's shoe was certainly based on reality; sharp footwear was developed by the KGB, though they may not have added poison. But adding a toxin taken from 'the sex glands of the Japanese globe fish', as we are told in the Fleming books, isn't that far-fetched.

The poison from the Japanese globe fish, or puffer fish, is tetrodotoxin, a powerful nerve toxin. The slow-swimming, vulnerable puffer fish accumulates and concentrates the poison from the bacteria in its environment to use for its own defence.[15] The toxin ends up in its skin, and some internal organs, which deters other animals from eating it. Another tetrodotoxin-accumulating animal is the blue-ringed octopus, the one in *Octopussy*. Octopussy's lethal pet looks impressive in its tank, but it is also useful when an attacker crashes through the glass and emerges with the animal wrapped round his head like a face-hugger.

---

[15] Other animals, including rough-skinned newts and moon snails, also take advantage of this poison-producing bacteria to obtain toxins for their own defensive or offensive purposes.

Just a few milligrams of tetrodotoxin can be fatal to humans because it disrupts normal nerve function. Electrical signals travel down nerve cells to pass on messages. Each cell is a tiny biological battery with positive and negative regions formed by accumulating potassium inside the cell and forcing sodium outside. To generate the electrical signal, channels open along the length of the nerve. Sodium rushes in through sodium channels and potassium rushes out through potassium channels. Tetrodotoxin blocks the sodium channels and the nerve becomes useless. Paralysis of the muscles, including those that allow us to breathe, is the result. There is no antidote. It doesn't look good for either Bond or Octopussy's attacker.

In *From Russia with Love*, Bond is saved from the fatal footwear when Klebb is shot dead by Tatiana. In the novel he is not so lucky. Klebb makes contact and Bond crashes to the floor. Readers had to wait another year for the publication of the next novel to find out if their hero had survived. Of course, he did. However bored Fleming had become with 007, he couldn't kill him off. Fans wrote concerned letters asking about the agent's health and Fleming was forced to resuscitate Bond for another instalment of adventures.

In the following novel, *Dr No*, he revealed that the dying Bond had been given mouth to mouth resuscitation, to keep oxygen entering his body, until a doctor arrived. He then spent a long time undergoing treatment for the effects of the poison, which in real life would be artificial support for breathing until the toxin is cleared from the body naturally. It is very effective.

MI6 got the Russian coding machine and Bond was able to return to work.[16]

The coding machine Bond went to so much trouble to obtain in *From Russia with Love* is clearly based on the German Enigma machine used to encode secret messages during the Second World War. These machines look much like a portable typewriter, but with a series of wheels where the paper might be expected to go. Each wheel has the letters of the alphabet around its rim. Underneath the keyboard are wires connecting one letter to another. As a message is typed out, each letter is transposed to another, through the wires, and then to another on the first wheel, and then another, and so on. Each key press moves the wheels onto a new position, so repeatedly pressing the same key produces a different letter each time. As the war progressed, more wheels were added to the Enigma devices, message length was restricted and many other developments were added to make the code more difficult to break. Billions of different combinations were possible.

So that the intended receiver was able to decipher the message, they had to know how the machine was initially set up. Getting hold of Enigma machines, codebooks and any other scrap of information about the device and its operation was a key goal for the intelligence services, but it was of limited use when the initial settings of these machines were changed daily. It took the combined efforts of brilliant mathematicians, skilful technicians building and maintaining computing machines to crunch

---

[16] It is doubtful the attacker in *Octopussy*, with the octopus stuck to his face, was so lucky.

through possibilities, with thousands more operating them, configuring their inputs, reading their outputs, deciphering and translating, to crack Enigma.[17] Knowledge of the UK's success in cracking the code was not made public until the 1970s, but Fleming was well aware of it and had made regular visits to the centre of the codebreaking activity, Bletchley Park, during the war.

As *From Russia with Love* shows, the gadgets of the Connery-era Bond films are mostly very practical, certainly within the world of espionage. They generally help Bond and his enemies achieve their goals more easily and in a range of situations. In fact, some were so good, the real secret services looked on in envy.

The re-breather[18] in *Thunderball* was a tiny tube that could be held in the mouth to allow a diver to breathe for four minutes underwater – enough time to see off a few henchmen with harpoons or enter a villain's lair via the underwater door. The Ministry of Defence was intrigued; it had been trying to develop something similar itself. Employees got in touch with the filmmakers to ask how they had done it, only to learn that the re-breather was just a couple of soda siphon capsules, and the actors were holding their breath.[19]

---

[17] The German designers of Enigma knew the code could be cracked but never thought anyone would go to the enormous effort needed to do it.

[18] Though it is usually referred to as a re-breather, it is actually a device to be used if a re-breather (a system that recycles a divers exhaled breath so no bubbles appear) wasn't available. It also reappeared in *Die Another Day*.

[19] Fiction has become fact and there are now genuine re-breathers available on the market. They look a little different to Bond's gadget

The radioactive tracking device Connery swallows in the same film, however, was fiction in 1965 – and still is. Radioactive tracers can be used to track down areas of medical interest inside the body. The dose is low and special detectors have to be placed close to the patient to get a reading. Swallowing something that can give off enough radioactive energy to be tracked from miles away would do horrendous damage to Bond's internal organs.

A more practical and conventional method of finding a lost or captured agent would be to use a radio transmitter. A small device emitting radio waves at a certain frequency can be tracked by a receiver, as shown in *Goldfinger* twice. One tracker the size of a packet of cigarettes is hidden in Goldfinger's car. Another is small enough to fit inside the heel of Bond's shoe, a trick repeated in *Skyfall*. The transmitter itself can be very small but it needs batteries to power it and these tend to be much bigger. The smallness of a tracker is limited by how long you want the signal to last and how far you want it to reach.[20] The Global Positioning System (GPS) enabled transmitters to be tracked anywhere around the world, but the size of these devices is still limited, by the antenna as well as the batteries.[21] The tiny trackers seen in 007 films aren't practical, but then Bond's gadgets aren't always about practicality, as Moore's Bond would demonstrate.

---

and work like fish gills, pulling in and concentrating oxygen from the water itself.

[20] In the Daniel Craig films these trackers have shrunk so small they can be injected into Bond's body (see chapter 025).

[21] Though the first GPS satellite was launched in 1978, a fully operational system of 24 satellites wasn't available until 1993.

## Toys for boys

Roger Moore's Bond has so many extras added into his watch it's surprising he isn't dragging his knuckles on the floor. In *Live and Let Die* his Rolex comes with an electromagnet powerful enough to deflect a bullet, but not powerful enough to pull a boat from its moorings.[22] The same watch also has a tiny buzzsaw that can cut through ropes. In later films his watch is upgraded with detonators, a glorified label maker so he can print the messages headquarters send him, then a two-way radio and even a TV.[23] While a lot of these additions are made for entertainment rather than to help Bond out of a fix, it is part of a trend in packing all manner of things into innocuous everyday items. The 007 films from Moore's tenure show darts, knockout gas, bullets, lasers, flamethrowers and acid released from watches, lipsticks, cigarettes, notebooks and pens.

Most of the time these gadgets utilise known technology that needs minimal explanation, but miniaturised. This miniaturisation may be beyond the technical capabilities of real life but not of the fictional Q-branch. Sometimes, however, it seems the filmmakers are inventing things they simply wish existed, like the X-ray specs given to Brosnan's Bond in *The World Is Not Enough*.

---

[22] Electromagnets are real but, long story short, one that powerful and that small would generate so much heat Bond's wrist would be cooked as soon as he switched it on.

[23] David Niven's Bond got his TV watch first, though, in 1967's *Casino Royale*. The film mercilessly parodies the abundance and absurdity of Bond gadgets long before Roger Moore was weighted down with so many of them.

The closest real-life equivalent of these glasses are either the joke-shop toys sold to kids, or the body scanners installed in airports. The scanners in airports actually work. The X-rays used in airport security do not penetrate as deeply as those used in hospitals to look for broken bones. These lower-energy X-rays are shone at passengers and some are reflected back onto a detector, to create an image. Some materials reflect the X-rays more easily than others and so clothes appear transparent,[24] but the body, and any contraband being smuggled, doesn't. Bond uses his glasses when he enters a casino to see who is carrying a concealed weapon, which is pretty much everyone, and something he could have guessed without Q-branch going to the trouble of packing an unfeasibly large amount of equipment into some very thin-framed shades. But it also means he and the audience can indulge in some schoolkid gawping at the female clientèle.[25]

The gadgets in Bond are a way of indulging a few fantasies, but of the technological kind. And they are not all as unbelievable as they might first appear, as their origins in the wartime secret services show. Regardless of whether a Geiger counter really could be squeezed into something as small as a watch in 1965 (*Thunderball*),[26]

---

[24] Including underwear. Only the film censors could stop everyone appearing naked through Bond's glasses.

[25] Why the women are wearing guns under tight-fitting dresses is a mystery. They would ruin the line of the dress and be near impossible to draw should they need to.

[26] Spoiler, it can't. But a Geiger-counter watch was released in 2003 by a Swiss manufacturer if you want to re-enact the *Thunderball* experience for yourself.

or a little platform that slides out from a laptop can do sophisticated materials and DNA analysis (*Spectre*),[27] what is genuinely ingenious is how all these gadgets, complex or simple, outrageous or practical, have been woven into the plots. *GoldenEye*, for example, would be a much shorter and very disappointing film had Bond not had a laser watch handy to get him out of a train that was about to self-destruct, but lasers are the subject of the next chapter.

---

[27] Again, no.

# 003

# *Goldfinger* and the laser

*Goldfinger* is widely regarded as the gold standard for the 007 film franchise. It has everything: a phenomenal title song, a great villain, a grand scheme and even grander sets and gadgets. This, for many, is what James Bond is all about. There are more memorable moments from this film than perhaps any other in the series, but one stands out among the others as encapsulating the very idea of a Bond film: the villain stood in an opulent set, nonchalantly looking down on a beaten and bound Bond, about to execute our hero in an elaborate and convoluted manner using the biggest, brashest gadget at his disposal. Auric Goldfinger plans to emasculate and execute James Bond with a laser beam between the legs.

The Goldfinger of Fleming's novel threatens to divide Bond in two with a circular saw. The idea was already fairly old hat even in 1959 when the book was published. To reflect the film's cutting-edge aspirations Bond's life, and masculinity, would be threatened by the latest technology. Goldfinger is given 'a scientific device so new that only a minority of the general public have even heard of it'. It was a fantastic choice.

Lairs, cars, guns, even diamond-studded satellites – everything is better with a laser attached. They are showy, futuristic and potentially lethal – every villain should have one. Smaller, stylish lasers discreetly hidden inside Bond's car or watch only enhance his sophistication.

These high-tech devices have become inextricably linked to Bond films.

When *Goldfinger* was released in cinemas in 1964 hardly anyone knew what a laser was or even looked like. Invented only four years before the film's release, lasers had not yet found many applications outside of niche scientific research. It meant the film's props department could get very creative without undermining the basic science of the laser itself. And, while lasers were certainly very modern, the idea of a glowing beam of light that could kill was already familiar. The death ray was a science fiction staple, and so well established in popular culture, audiences wouldn't need lengthy explanations that would slow the pace of the film.

## Death rays

The idea of using light as a weapon is very old. It started with Archimedes, the Greek polymath. In 212 BC the coastal city of Syracuse was under siege, its harbour blockaded by Roman ships. Knowing that a curved reflective surface focused light and could be used to ignite kindling, Archimedes proposed installing huge, curved sheets of metal along the city walls to concentrate the rays of the midday sun onto the enemy's wooden ships, setting them alight. It might have worked, but the mirrors were never made. It's almost the same idea used by Gustav Graves in *Die Another Day*, but instead of curved metal sheets on harbour walls, he has a giant parabolic mirror in space that can focus a destructive beam of light on a place of Graves' choosing.

A giant space mirror might seem one of the sillier additions to *Die Another Day* – and it's a crowded field – but it is not completely ridiculous. In the 1990s Russia launched Project Znamya, a satellite with a 20m (65ft) wide solar mirror to reflect the sun's rays onto regions where winter days were particularly short. The first satellite reflected a 5km (3 mile) wide spot of light, as bright as a full moon, on the ground below. A second 25m (80ft) wide mirror was launched, with the aim of generating a 7km (4.3 mile) wide spotlight with the intensity of five to ten full moons. But the mirror got caught on an antenna and ripped. The project was abandoned.[1]

Graves' space weapon is big, bright and very destructive, but it isn't a laser. The column of light from Graves' Icarus satellite is a concentrated beam of natural sunlight. A laser, as Goldfinger so eloquently put it in 1964, is 'an extraordinary light not to be found in nature'. What, you might ask, is the difference?

Light, made up of packets of energy called photons, is produced when an electron drops from a high energy state to a lower energy state. The heat of the sun, the electricity flowing through a tungsten wire in an incandescent light bulb and chemical reactions are just some of the ways energy can be provided to make an electron jump to a higher energy level. When an electron drops back to a lower energy level it releases the energy it absorbed. If the gap between the higher and lower energy levels is within a certain range of values we will perceive the energy released as light. This light can have a range of wavelengths, or colours, and shine out in all directions. It's a little like a fountain.

---

[1] Even if it had worked, it couldn't have produced the intensity and manoeuvrability Graves needs for his evil plan.

Water, or electrons, are constantly being pumped up higher and falling back down.

A laser is not like a fountain, but more like a coin-drop game in an arcade. Laser is an acronym for 'light amplification by the stimulated emission of radiation'. What that means is light is used to pump electrons in a specific substance, called the lasing medium, to a higher energy level, where they get temporarily stuck. So more and more electrons, or coins, are added to the higher level. Then there is a trigger, a photon, or one coin too many, that sends a few falling to a lower level. One coin falling can drag more with it, and one photon can trigger more photons to be released, which trigger yet more, and the light is amplified. The photons that are released, as the electrons fall, all have the same energy or wavelength because they all fall from one specific energy level to another specific energy level – like the coins falling from one shelf to the next. All the photons travel in the same direction. Lenses and mirrors can be used to focus this stream of photons into an intense, coherent beam – like columns of soldiers marching in a straight line.

## A thoroughly modern weapon

Theodore Harold Maiman built the first machine capable of producing a laser light in 1960.[2] In true Bond-villain style, his device used a precious stone, a ruby, to generate a beam of red light that became the standard, almost stereotype, of any laser depicted on screen. *Goldfinger* was the first film to feature a laser, so those watching when it was first released didn't have the

---

[2] Based on theoretical work done by several others.

benefit of any prior knowledge to help them understand what they were looking at. Fortunately, Bond villains are always happy to talk at length, not just about the magnificence of their intellect and the scale of their plans, but also the ingenious way in which they are going to kill Bond and the technological superiority of the device they will use to do it.

Goldfinger explains to a bemused Bond that the device pointing at his privates is 'an industrial laser'. Delighted with his toy, he continues: 'It can project a spot on the moon or at closer range cut through metal. I will show you.' And he does: a red beam appears and starts to slice through the sheet of gold Bond has been strapped to.

With the power of the laser amply demonstrated, it would be a simple job to use it to kill Bond very quickly. Simply point the beam towards the head or heart and the damage caused would result in a rapid death. But why would a Bond villain as extravagant as Auric Goldfinger go for the quick and efficient method when he can be elaborate and dramatic?[3] The laser beam continues its slow progress through the metal sheet, between Bond's spread legs.

If Goldfinger's claims are true, that his laser is powerful enough to cut through metal, it will make short work of slicing through a secret agent. Had Goldfinger decided not to spare Bond's life, it would be a very unpleasant end. The laser would have started its bisection at one of the most sensitive parts of Bond's anatomy and it would have to work its way through quite a bit of his body

_____

[3] He also has ranks of minions who can clear up the mess for him afterwards.

before it reached vital organs.[4] The beam would also pass through a few major blood vessels on its way and, though the heat from the beam may cauterise tissues as it passes through,[5] it is unlikely to be enough to staunch the loss of blood completely. Hundreds of blood vessels that were once connected to allow blood to circulate throughout Bond's body would be cut off. When there is insufficient blood moving through the circulatory system to deliver oxygen, it is referred to as 'shock'. After the first few pints have been lost, breathing becomes heavy and rapid as the body tries to compensate for the lack of oxygen normally being transported to tissues around the body. The heart speeds up for the same reason. Eventually blood pressure will drop to a point where the heart could no longer keep pumping it. When the heart itself isn't receiving enough oxygenated blood to perform normally it slows, until it eventually stops. Bond is likely to die of circulatory collapse long before the slow-moving beam could reach his heart. Thankfully, the body can be very good at blocking out the worst of the pain it is experiencing (see chapter 021). Bond would probably be unconscious relatively early in the process.

Lasers certainly can be used to cut through metal and human tissue, just not the one we see on screen. The red light from Goldfinger's laser would be almost completely reflected by the metal and it doesn't have the power to cut through Bond's body either. Infra-red light, generated by carbon dioxide lasers, is what you need, but this light is beyond our visual range. An invisible laser beam would

---

[4] Your definition of a vital organ may differ.
[5] This is just one of the many advantages of using lasers in surgery.

be confusing, so a bit of artistic licence is needed for the audience to know what is going on.[6]

The Bond filmmakers often go to extraordinary lengths for authenticity, and they did have a real laser brought into the studio, but the cameras simply couldn't pick up its red beam. This is not surprising. We can't see light itself, only when it interacts with matter. For example, the light from laser pointers used in presentations, or the laser sights that have featured in so many Bond films, can't be seen along the entire length of the beam, only where it bounces off the presentation screen or Bond's suit to show that he is being targeted. The bright red beam in *Goldfinger* was added in post-production to literally highlight what is happening. The continuous line of red light could be explained away by Goldfinger's gold foundry being particularly dusty, with the beam bouncing off particles in the air.

In short, James Bond was in no danger of being harmed by Goldfinger's laser beam, but Sean Connery definitely had something to worry about. To show the metal sheet apparently being cut in two by the beam, Burt Luxford, special effects technician, was under the table with a blowtorch, slowly burning his way through a line of solder. Luxford couldn't see where Connery started or stopped so had to wait for a particular line of dialogue to turn off the torch. Gert Frobe in the role of Goldfinger, as every Bond villain should, dragged out his lines as long as he could.

Goldfinger established a firm link between advanced technology, in the form of lasers, and their misuse by

---

[6] There is also the small detail that metal-cutting and surgical lasers hadn't been invented yet.

master criminals. The same technology can, of course, be used for the good. But Bond wouldn't get his hands on a laser until 15 years and eight films later.

## Lasers for everyone

The 1979 film *Moonraker* launched Bond and CIA agent Holly Goodhead into space on a mission to thwart Hugo Drax's plans to destroy humanity. Everyone knows that adventures in outer space involve laser guns and when it comes to equipping Bond for his mission, Q doesn't disappoint. Q-branch has developed laser weapons and thoughtfully shared the design with the Americans. When it comes to the final big space battle, Drax's minions have their laser weapon built into the chest piece of their spacesuits. The US space marines have theirs in an over-the-shoulder design. On board Drax's space lair, the fighting continues with laser pistols for the minions and bigger laser shotguns for the good guys. Yet more laser weapons have been built into Drax's space shuttles, which is handy for Bond, who uses one, and his excellent shooting skills, to destroy the globes of poison that are heading towards Earth.

The space shuttles and the astronauts' spacesuits were modelled on those being tested by NASA at the time of filming. The weapons they are armed with, however, are more closely modelled on those of science fiction.[7] These guns are clearly lasers guns but they emit a blue-green beam rather than the traditional red. Lasers come in all different hues depending on the lasing medium being used

---

[7] *Moonraker* was very much cashing in on the new vogue for space-based films, thanks to the enormous success of *Star Wars*.

to generate the photons in its beam. Blue and ultraviolet lasers are more difficult to produce because these photons have more energy, but ultimately it is a question of finding the right material, with the right gap between its electron's energy levels to releases the right photons. Blue-green lasers suggests Drax and the Americans have gone for more powerful weapons, but the colour may have been a simple practical choice by the filmmakers so it would show up better against the black background of space.

No matter what colour of laser light is being generated, the beams, just like those of Goldfinger's metal-cutting laser, shouldn't be visible at all. The vacuum of space won't even have dust floating around, as there could be in Goldfinger's lair, to show us the path of these lasers as they head towards their targets. It makes no sense scientifically but it is entirely sensible from a filmmaker's point of view. Having lasers that you couldn't see until bright spots appeared on their target would be confusing, and the huge space battle would make no sense to viewers.

With allowances for a bit of artistic and dramatic licence, the weapons themselves are also shown to be effective. It is not unreasonable to assume that powerful, pulsed blue-green lasers could burn their way through the layers of a spacesuit, creating a hole through which the air inside can escape. The occupant of the spacesuit would likely be killed by asphyxiation (see chapter 020) rather than by damage from the laser itself.

Q-branch may have been a little slow on the uptake, but *Moonraker* showed that lasers could be an extremely useful tool and not just the province of megalomaniac Bond villains. While the villains would continue to go in for big, flashy laser weapons in keeping with their egos, the lasers designed for Bond's use are sleek, efficient and unassuming, to match his professional status as a secret agent.

## Size doesn't matter, it's what you do with it

*The Living Daylights* introduced a new Bond, Timothy Dalton, but brought back many of the traditional elements that had been missing from Moore's tenure in the role. The Aston Martin returns, equipped with lasers in the hub caps in traditional red colour.[8] The car and its driver are spotted heading towards the Austrian border by the Czechoslovakian police. The police attempt to pursue the V8 in their Lada[9] and Bond considerately slows down to allow them to catch up. It is not, however, the gentlemanly, law-abiding gesture it appears. As the Lada slowly edges past Bond's car he uses his car's laser beam to cut through the police car's bodywork. When the police officer slams on the brakes, the car separates. The vehicle's momentum pushes the body of the car forward, complete with the two police officers inside it, and the wheels and wheelbase are left behind. It makes no sense, but that isn't the point. Bond has thrown off his pursuers without hurting anyone. Bond only kills bad people, and only when he has to – never innocent bystanders, or even police officers just doing their job. Property, on the other hand, is fair game. Bond has turned the destruction of vehicles, buildings and any other object in his path into an art form.

---

[8] Q-branch also found space inside the V8 Volante for not just lasers but every conceivable gadget, all of which get used in one chase sequence before Bond lands the car in a snow drift.

[9] The Lada VAZ-2103, I believe. In the Soviet Union they were called Zhiguli, which apparently sounded too similar to 'gigolo' when said in an English accent, though seems appropriate for a Bond film.

In 1987, Q-branch had managed to scale down the size of a metal-cutting laser to fit inside a car wheel. By 1995 and *GoldenEye*, they had shrunk the hardware even further to fit into a wristwatch.[10] The thin red beam emerging from Bond's Omega watch can apparently cut through 2.5cm (1in) armour plating in seconds. The same watch-based gadget makes a reappearance in 2002's *Die Another Day*, but this time all it has to cut through is ice. Admittedly it is quite thick ice, but still ice that can melt at 0°C (32°F) and not steel that melts at over 1,400°C (2,550°F).

Lasers can certainly be small – just look at the tiny pen lasers used in presentations. And, while these mini lasers might cause problems if shone directly into someone's eyes, they are not up to cutting through steel. Several people have tried to build themselves a Bond-style laser watch for real and there are many videos demonstrating that it is possible to increase the power enough to ignite a match with their beam. It's impressive, but not nearly powerful enough for Bond's purposes.[11] Using a watch-sized laser to melt ice rather than metal is more credible, but a foot of ice would take some patience to get through. However, Bond has the technical support of Q-branch behind him and his gadgets. They obviously have access to the best technology, and they also have Q in charge, so almost anything may be possible (see chapter 002).

---

[10] The laser watch was actually first seen in one of the few Bond films that is not part of the main franchise, the 1983 *Never Say Never Again*.

[11] It would also seem easier to use the red strip on the side of the matchbox to light the match, but this is missing the point.

**Fatal status symbols**

Over the course of the 007 franchise, lasers have
developed from an interesting novelty item to being so
commonplace they are practically part of the furniture.
In *Die Another Day*, we know Graves must be a villain
because his lair comes complete with an industrial laser
even though he has no obvious use for one. In a deliberate
homage to the scene in *Goldfinger*, the baddies find a use
for this expensive accessory by threatening to cut off
Agent Jinx's head with its red beam.

The 007 films often exaggerate the capabilities of the
lasers they include, but in some cases real-life science and
technology has caught up. In 2002 lasers certainly were
capable of cutting through a person's neck (just not red
ones). Fortunately, Jinx escapes and it is the henchman,
Mr Kil, who gets the full force of the beam through the
back of his skull and brainstem, obliterating basic functions
and bringing about his rapid demise. By contrast, the
biggest laser, the diamond-studded space laser, wielded by
the baddest Bond villain, Blofeld in 1970's *Diamonds Are
Forever*, is still a little way off, technologically speaking (see
chapter 007). After all, a Bond film isn't a Bond film if it
isn't at least a little over-ambitious.

Lasers are the perfect illustration of the franchise's
obsession with science and technology. They are
simultaneously traditional, in that they have been part of
the series almost from the very beginning, but are also
cutting edge, because they can be upgraded, miniaturised
and adapted to the needs of the story. Regardless of how
realistic these situations really are, lasers will forever be
associated with the 007 films.

# 004

# *Thunderball* and the gamma gas

Bond villains have a lot riding on their grand schemes. They are also particularly intolerant of anyone who does not wholeheartedly help them accumulate the personal wealth or power they crave. Their complex global plans involve thousands of people, any one of whom could throw a literal or figurative spanner in the works. Dissenters, nosy agents, even those who want too big a slice of the villain's pie, all must be swiftly and discreetly dispatched so as not to rouse suspicions before their plans can be put into action.

History, and fiction, is littered with the bodies of those who got in the way of someone's ambition. Rulers, influential people, the wealthy and those who simply knew too much, all have been targets for assassination for thousands of years. Though many methods have been used, poison has always been a firm favourite. As time and technology has moved on, the poisons have been updated but the method has never gone out of fashion.

The 007 franchise has seen dozens of inconvenient individuals stabbed, shot, electrocuted and have their neck broken with weaponised bowler hats, to name but a few personalised methods of execution. But sometimes the old ways are the best ways. If poison was good enough to kill emperors and kings, it is good enough for a Bond villain. In a franchise that is well known, almost celebrated, for its far-fetched means of killing people, some of those chemical killings are uncomfortably close to reality.

## Gamma gas

In 1965's *Thunderball*, Emilio Largo, a member of
SPECTRE, has hatched a plan to hijack a military plane
carrying two nuclear warheads, which will then be used
to hold the world to ransom.[1] Months have been spent
putting a SPECTRE agent through extensive plastic
surgery to make him look like Major François Derval,
the pilot who will be flying the plane. The real Derval
needs to be removed, permanently and without any fuss,
if their plan is to succeed. A quick shove into a shark
pool would get rid of both him and the evidence, but
Largo's shark pool is in the Bahamas, and the flight is set
to take off from the UK.[2] Fortunately for Largo, and
unfortunately for Derval, the hijack plan involves the use
of a poisonous gas to kill the rest of the crew on board
the flight. A test squirt of the same gas in Derval's face
will simultaneously check its effectiveness and get rid of
someone who could mess up the plan.

In the novel, straightforward cyanide (see chapter 023)
is used to kill off the crew aboard the hijacked flight.[3] In
the film, we are told it is 'gamma gas'. A quick spray to
the face, Derval coughs a bit and collapses dead on the
floor. Such a rapid collapse would be easily believable

---

[1] It is the same basic plot as the non-EON Bond film *Never Say
Never Again* – the result of long, complicated legal wranglings that
gave Kevin McClory the right to film his own version. (EON is the
film production company formed by Cubby Broccoli and Harry
Saltzman).

[2] Shark pools are rather thin on the ground in rural England.

[3] The pilot is also in the pay of SPECTRE and so there is no need
for the double, or the cosmetic surgery, and they only need to kill
him after he has completed the mission – easily done by drowning.

had Fleming's original cyanide been used, but there is a problem. There are two other people in the room with Derval: the man who sprays the poison and Derval's girlfriend.[4] Neither are affected by the poison, even though they are standing just a few feet away and doing nothing more to protect themselves than covering their faces with a piece of cloth. A small window in the tiny room is opened and this is apparently enough to allow the gas to escape, or be diluted to sufficiently low levels, that the other two can dispense with their thin face coverings and continue their conversation with no ill effects. Gamma gas is clearly not cyanide.

It would be understandable if a few scientific corners had been cut to keep the film moving along. But, however improbable the scenario may seem, it is not as far-fetched as you might think.

On 13 February 2017, Kim Jong-nam, half-brother of the North Korean leader Kim Jong-un, was at Kuala Lumpur International Airport waiting to catch a flight to Macau. He was approached separately by two women. One sprayed a liquid in his face. The other put a liquid-soaked cloth to his face. He walked to a reception desk to report what had happened and a short while later he collapsed. Kim was rushed to hospital, where he was treated with atropine and adrenaline and given a tracheotomy to help him breath. He died later the same day. An autopsy confirmed poisoning and VX nerve agent was the prime suspect.

---

[4] Spoiler, she is a SPECTRE agent, and only with Derval so she can kill him.

What puzzled many people was how Kim had been able to walk calmly over to the reception desk without showing any obvious signs of poisoning. Also, those around him were apparently unaffected.[5] The area where the incident occurred wasn't cleaned for several days but there were no further reports of people falling ill. This should not have been possible if it really was VX that had been used in the attack.

VX is a potent nerve agent developed by the British in the 1950s and is part of a very large group known as organophosphorus compounds. These compounds were initially developed as insecticides until their toxicity to humans, as well as insects, became apparent. The degree to which any particular organophosphorus compound affects the human body varies, meaning some of them are used as commercial insecticides while others are so deadly there are international treaties banning their use in any circumstances. Several organophosphorus compounds have become notorious through their use in war and terror attacks, for example Sarin, used in the Tokyo subway in 1995, killing 12 and severely injuring 50 more.[6]

Organophosphorus compounds disrupt normal nerve function in insects and/or humans. Messages are passed between nerves by releasing chemicals from one nerve, which dock in receptors on a neighbouring nerve. Different chemical signals are used to send different messages and to different nerves. Acetylcholine is the chemical that

---

[5] One of the women involved vomited in her taxi leaving the airport and reports feeling unwell ever since.
[6] This real-life incident was referenced in the film *Tomorrow Never Dies*.

tells the parasympathetic nerves – those responsible for unconscious activities in the body such as breathing, sweating and digesting – to activate. All the time acetylcholine molecules are docked in the receptors, the nerve is receiving the 'go' message. An enzyme, cholinesterase, breaks down the acetylcholine molecule to stop the message, allowing the nerve to reset so it can receive new messages in the future. Some organophosphorus compounds can block the cholinesterase enzyme, meaning the acetylcholine cannot be removed from the receptors and the parasympathetic nervous system goes into overdrive.

The symptoms of organophosphorus poisoning are a consequence of the body's 'rest and digest' processes going haywire. There will be profuse sweating, stomach cramps, involuntary urination and defecation, the pupils contract to pinpoints and everything appears to go dark. The heart rate will slow and the victim will struggle to breathe.

As little as 10mg of VX, less than a grain of rice, can kill if it comes into contact with the skin. The amount needed to kill plummets to a terrifyingly tiny 5µg (a dose thousands of times smaller) if it is inhaled. VX is actually a liquid, not a gas (its boiling point is over 300°C, or 572°F).[7] Spraying the liquid in an aerosol would be an effective way of delivering deadly amounts onto the skin or into the lungs. The aerosol is heavier than air and will not reach other people in the room so

---

[7] Perhaps this is the liquid dripped down the string in *You Only Live Twice*, intended for Bond but it ends up dropping into the mouth of agent Aki. It kills her in a matter of seconds, after a few gasps and convulsions.

easily and a cloth over the mouth, though not a perfect barrier, might help trap at least some of the aerosol droplets. Even if some of the droplets made it onto the assassin's skin, they could protect themselves from its effects with an antidote.

The symptoms of organophosphorus poisoning can be reversed by promoting the opposite response, increasing the heartbeat, drying up secretions and dilating the pupils.[8] This can be done very effectively using atropine, which blocks acetylcholine receptors and inactivates them.[9] There is no sign of antidotes being used in *Thunderball* but there are more reasons why VX is still a good candidate for gamma gas.

VX can also be used as a so-called 'binary' weapon. Two relatively non-toxic chemical precursors can be combined to produce VX in situ. For example, a woman can approach a man in an airport and spray one precursor in his face, and a second woman could then wipe another liquid onto the face. A chemical reaction would then occur between the two liquids to make VX on the man's face. The reaction may not be perfect but enough VX could be produced to cause death while leaving very few, if any, others affected. It would explain how Kim Jong-nam was killed but the two women who attacked him were not.[10]

---

[8] The so-called fight-or-flight response (see chapter 021).

[9] Soldiers deployed in areas where they might encounter these kinds of nerve agents carry epipens, or autoinjectors, pre-loaded with atropine so they can treat themselves immediately, avoiding the worst effects until medical help can arrive.

[10] It also means the two women may not have been aware of what they were involved with and genuinely thought they were acting out a simple prank with innocuous substances.

It would also explain how everyone in the hotel room in *Thunderball* is fine except for poor Derval. Something could have been smeared on his face beforehand, perhaps in aftershave or from a towel or pillow. Unfortunately, in the film we only see one substance, the spray.

Organophosphorus compounds can be modified and manipulated by scientists to optimise their properties for the required purpose. They offer a lot of choice to the potential poisoner, but there are plenty more toxic substances that can be, and have been, used for assassinations in both Bond's world and the real one.

## Spy fishing

*A View to a Kill* sees Bond in Paris, seeking information about Max Zorin, megalomaniac for this particular instalment of the franchise. Over lunch in a restaurant at the top of the Eiffel Tower, he asks the French detective, Monsieur Aubergine, what he knows.[11] Aubergine clearly has his suspicions about Zorin's activities but, before he can elaborate, a butterfly act appears on the stage to entertain the diners. Paper butterflies flit around the room, but one, a bright yellow and black butterfly, flies around wildly and slams into Aubergine's cheek. He promptly falls dead in his dinner.

Whatever poison the fake butterfly was carrying, it was fast-acting. The substance is never identified but, as we learn later, the killer has access to a wide range of chemicals from the lab where Zorin is developing drugs to dope his racehorses and perhaps humans too. Maybe one of these drugs, given in a sufficiently high dose,

---

[11] An easily missed pastiche of Hercule Poirot.

could have caused Aubergine's death. Administering the poison, however, can be more tricky. A poison-laced barb on a fake butterfly on the end of a fishing rod sounds ridiculous, but it puts a bit of distance between the assassin and their target. Hiding in the shadows with the other puppeteers makes the killer more difficult to spot and the surprise of the unusual method gives them a bit of extra time to escape. If you want to get away with murder it can be worth taking the time to form an elaborate and unusual plan,[12] as the real-life case of Georgi Markov shows.

In 1978 Markov, a Bulgarian defector, was working in London as a journalist. On 7 September he took his usual route into work and while standing at a bus stop on Waterloo Bridge he felt a sting in the back of his leg. He turned to see a man pick up an umbrella, apologise, cross the road and get into a taxi, never to be seen again. Four days later Markov was dead.

An autopsy revealed a tiny pellet embedded in his thigh below the spot where he had felt the sting. The pellet was tiny, just 1.5mm (0.06in) in diameter and had two holes drilled into it. Where the two holes met, a cavity was formed, which analysts guessed had been filled with ricin.[13] It was hypothesised the pellet had then been coated in either wax that would melt, or sugar that would dissolve, to release the poison into the body. To force the pellet under the skin without damaging the poison would require some kind of adapted air pistol. One theory is

---

[12] Better still, don't kill people.

[13] Ricin and its properties will be discussed in more detail in chapter 006, *On Her Majesty's Secret Service*.

that the umbrella Markov noticed was in fact modified to act as a rifle. Another theory has a more conventional air pistol being used but the umbrella dropped as a distraction. The truth will probably never be known.[14]

Markov's death shocked the world for many reasons. Secret agents were said to be involved, it was brazenly carried out in a very public spot and using a relatively unknown poison (for the time) delivered by a specially adapted gadget. It was like something out of a Bond film, and *A View to a Kill* seven years later proved it. Other Bond films have gone for more straightforward methods of poisoning.

### Drink responsibly

*Casino Royale* sees Bond pitted against Le Chiffre in a high-stakes game of poker. Le Chiffre decides to tip the odds in his favour by killing his opposition with poison dropped in Bond's trademark cocktail.[15] It's like an Agatha Christie plot with a high-tech twist. Bond takes a sip and soon starts to feel unwell. Unsure of what he has swallowed, he excuses himself to stagger off and try to counteract its effects.

First he stumbles into the bathroom with a glass and a salt cellar. He swallows a saltwater slurry to make him vomit out as much of the poison as possible. Preventing more poison being absorbed into the body proper is an excellent strategy, but vomiting can make things worse.

---

[14] No one has ever been arrested let alone faced trial for Markov's murder.

[15] In the novel he is threatened with a gun pressed in the small of his back by one of Le Chiffre's henchmen.

If you have swallowed something harmful, call the emergency services and let them treat you. Bond has access to a very personalised emergency service courtesy of MI6. His company car is kitted out with a mini medical diagnostics kit and a defibrillator, presumably standard issue for secret agents in case their health is compromised.

As Bond's symptoms worsen, he sways out of the casino, sweating profusely and struggling to focus. He gets to his car and plugs himself into the medical system. His results are relayed to London, where doctors are on hand to interpret the data. The diagnosis is poisoning by digitalis, a name many will be familiar with from commonly used heart medication.

Digitalis is the name given to a group of compounds found in foxgloves (plants of the genus *Digitalis*). Several of these compounds, notably digitoxin and digoxin, have a very strong effect on an important enzyme called $Na^+$/$K^+$-ATPase. This enzyme is a kind of molecular pump moving sodium ($Na^+$) and potassium ($K^+$) in and out of nerve cells to create the electrical current that transmits a signal along the length of the nerve. Variations of the enzyme are used in different types of nerve in different parts of the body. Digitalis compounds interact predominantly with nerves in the heart, hence their use in heart medicine.

Digoxin and digitoxin can help coordinate the nerve signals needed to regulate the heartbeat and also make the heart cells contract more strongly. With the right dose, weak, fluttering, uncoordinated heartbeats become strong, regulated contractions that push blood round the body efficiently. But too much digoxin and the nerves start to fire erratically and the contractions can become so intense that the heart is effectively paralysed.

Interactions with similar enzymes in other parts of the body are much weaker but can cause side-effects in a few people. Most commonly this results in nausea or loss of appetite, but in a few rare cases the $Na^+/K^+$-ATPase enzymes in the eye are affected. Often reported are changes in colour perception – everything appears more yellow. For some everything appears hazy, as it seems to do for Bond.[16] The symptoms produced from these side interactions are more likely to occur if there is a lot of digitalis in the body, as is the case with 007. The camera losing focus as he stumbles to his car is a neat way of suggesting his eyesight has been affected by the drug.

It is rare for cinema to accurately portray the effects of poisoning, but completely understandable when the symptoms can be very unpleasant (vomiting, convulsions etc.), and they usually continue for a long time – not something a viewer would want to linger over. To its credit, *Casino Royale* shows a fairly accurate depiction of digitalis poisoning but at an understandably accelerated rate for the benefit of the pacing of the film.

And what of the treatment that saves Bond so he can go back to the poker game? He injects himself with novocaine,[17] which is certainly a recommended first response treatment for a digitalis overdose. But it isn't enough for Bond. The heart monitor in *Casino Royale* shows Bond's heart rate increasing, which is incorrect

---

[16] More rarely, halos of colour appear around points of light.
[17] The same stuff used by dentists to numb the mouth before drilling.

for digitalis poisoning.[18] A rapid, uncoordinated heartbeat is known as fibrillation and can be effectively treated using a defibrillator, as Bond tries to do. He collapses before he can administer the shock to his heart and Vesper must do it for him. This is dangerous for digitalis poisoning because the defibrillator could cause further, potentially lethal disruption to Bond's already abnormal heartbeat. Of course, there is still the best part of an hour to go in the film so Bond survives.

Dropping something into a spy's drink may seem a little old hat, but it fits in with the back-to-basics feel of the Daniel Craig-era Bond. It also had some very uncomfortable parallels with real-life events that were going on in the UK around the time of the film's release.

On 1 November 2006 Alexander Litvinenko, a Russian defector and former member of the FSB secret services, was admitted to hospital. He was clearly unwell but doctors struggled to diagnose the cause. He suspected he had been poisoned, but he didn't know what with, and accused the Russian government of being behind it and several other attempts on his life. As several newspapers noted at the time, it was like something out of a Bond film.

Before he fell ill Litvinenko had met with two Russian contacts in the bar of a London hotel. A pot of green tea was already waiting for him when he got there.[19] Unknown to him, the tea had been laced with

---

[18] In *No Time to Die*, Safin comments that the foxgloves on Swan's desk can be dangerous if you eat them because they can make your heart stop – correcting the mistake from *Casino Royale*.

[19] Is Madeleine Swan being given a cup of green tea by a self-confessed poison expert in *No Time to Die* a reference to this case or am I overthinking things again?

polonium-210, a substance very few people were familiar with and which, as far as anyone knew at the time, had never been used to poison someone.

As Litvinenko's health deteriorated, more symptoms appeared, offering medics more clues. At one point his hair started to fall out. This is a classic sign of thallium poisoning and, combined with other symptoms, it was thought it might be a radioactive form of the element. Tests revealed slightly elevated thallium levels in his body, evidence perhaps of a previous attempt at poisoning, but not enough to explain his current situation. Poisoning by polonium-210 was only finally confirmed the day Litvinenko died, using a special test that had to be devised from scratch. It had taken 23 days for the radiation emitted by the polonium to destroy his organs. Even if the identity of the poison had been determined earlier there was nothing that could have been done to save him.

*Casino Royale* was released in cinemas on 17 November. Audiences watched Bond struggling against the effects of a poison dropped in his drink the same month Litvinenko was in hospital struggling for his life. It was an extraordinary coincidence. Fiction had unwittingly mimicked real life. And the real-life Litvinenko case provided some of the background inspiration for a later instalment of the fictional adventures of 007.

**The poisoner's poison**
Litvinenko wasn't poisoned by thallium, but it was not an unreasonable guess by the medical staff treating him. Thallium can disrupt several biological functions because the body mistakes it for potassium, an element involved

in many vital processes from nerve function to protein folding. Any and all of these functions can be disrupted by the presence of thallium and the result is a myriad of symptoms that are often misdiagnosed. One common symptom, however, is hair loss. But hair loss can also be caused by other things, including the radiation from polonium-210, as shown by Litvinenko.

Though not as well known as other poisons like arsenic, thallium has a nasty reputation. It was one of the many methods considered by the CIA to assassinate the Cuban leader Fidel Castro. The plan was to add thallium to Castro's shoes so it would be slowly absorbed through his skin, poisoning the leader but with symptoms that would be diagnosed as something else. The fact that it would cause Castro's iconic beard to fall out was considered a bonus.

The combination of a substance known as the 'poisoner's poison' and its association with the world of espionage made it an obvious choice for a Bond plot. Choosing a radioactive form of thallium gave the poison a Bondian twist and even more potential to do harm.

Thallium has many isotopes, atoms with a different number of neutrons. They are still thallium atoms, and behave like thallium, but they are not always stable and decay into more stable forms by spitting out energy and particles. The energy and particles ejected in the decay process can do damage by smashing apart molecules in the body. In practical terms for poisoning, only one isotope survives long enough to be produced, transported, administered and absorbed into the body. This is thallium-204, which emits beta radiation – dangerous certainly, but not as destructive as the alpha radiation emitted by the polonium–210 given to Litvinenko.

Radioactive thallium makes its appearance in the 2015 film *Spectre*. Bond is again in search of information and visits Mr White to get it. He finds him in a remote mountain hideout, alone and looking very, very ill. White explains his poor health is due to the radioactive thallium he found on his mobile phone.[20] So the poison must have been absorbed through his skin. This is certainly possible, but the lethal dose of thallium is about a gram, and that's a lot to have dusted over a phone without White noticing and wiping it off. Perhaps, unknown to White, extra thallium has been soaked into his clothing or shoes to ensure a lethal dose, because he is clearly ill.

He says he has a few weeks, maybe less, to live but doesn't show the classic symptom of thallium poisoning, or radiation poisoning, because he has a full head of hair and a beard. Maybe the full effects have yet to take hold. White also believes there is no antidote, but this isn't true.

Thallium poisoning can be difficult to diagnose but it is relatively simple to treat. Prussian Blue, the pigment used in paints and photography, can trap thallium within its structure, allowing it to be flushed out of the body. Radioactive thallium is chemically no different to non-radioactive thallium and would be removed from the body in the same way. Seeking proper medical treatment

---

[20] I'm not going to ask how he figured this out. His technologically well-appointed cabin in the woods perhaps has an atomic absorption spectrometer and a Geiger counter alongside all that surveillance equipment. What villain's lair wouldn't?

in a hospital may expose White to more risk; there are people who obviously want him dead and are prepared to go to considerable lengths to achieve it. And White may be stuck in a remote location, but Prussian Blue is very easy to obtain.[21] If he really has been poisoned with thallium, it is not necessarily a death sentence.

Poisons are the perfect fit for the vengeful, dangerous and strange world of James Bond. Their unusual properties, scope for creative methods of administration and long history of bizarre and sinister deaths fit right into 007's fantastical plots. Real or invented, the franchise has barely scraped the surface of possible chemical killings.

---

[21] Though not medical grade or formulated as an antidote, major online retailers, with efficient delivery services, carry stocks of this stuff for artists and photographers. But please don't self-medicate.

# 005

# *You Only Live Twice* and the volcano lair

It is a truth universally acknowledged that a Bond villain in possession of a plan for world domination must be in want of a lair. That lair must be befitting of its owner – larger than life, high maintenance and hugely impractical. Dictators, baddies and criminals, in real life and fiction, have always had their hideouts/castles/palaces/bunkers/ secret headquarters, but none of them can hope to compete with the scale and ambition of a Bond villain.

Bond bad guys have brought remote working to another level. The depths of the Amazonian jungle, the edge of the Arctic Circle or the vast emptiness of outer space – nowhere is too inaccessible for them to construct their lair.[1] As well as needing space for a helipad, nuclear reactor, monorail, shark pool, holding cells, weapons stores and accommodation for their huge staff of technicians, henchmen and minions, these megalomaniacs must build on the scale of their egos. Their bespoke bases of operation are vast, complex and with high design standards. They're not going to be satisfied with fixing up a simple two-bed semi in the suburbs.

Ken Adam's designs for Dr No's swanky dining room, sleek nuclear power plant and minimalist intimidation

---

[1] Of course, there are offices in the heart of Paris and Rome for regular meetings.

room/spider store brought the admiration of many.[2] Audiences gasped at his designs for Goldfinger's sophisticated Kentucky stud farm/operations room and the stunning interiors for Fort Knox. But they fell over backwards trying to take in the enormity of Blofeld's base in *You Only Live Twice*. The fifth film in the 007 franchise set the bar impossibly high. The hollowed-out crater of a volcano, with launch pad for rockets, retractable roof and working monorail, has been referenced, satirised and recreated in many films, cartoons and TV shows, but it has never been bettered. Even the 007 production team has struggled to live up to its own standards, creating ever more outlandish installations for its villains to carry out their evil plans.

If you have the unlimited resources of a Bond villain, why wouldn't you construct a lair to satisfy the size of your self-importance? Why not build your base in a giant satellite dish hidden under a lake (*GoldenEye*)? Or in a large underground cave system accessed via graveyard (*Live and Let Die*)? Need a bit more space? How about running your operation out of an abandoned island city (*Skyfall*)? With some basic infrastructure already in place, it's almost practical, although it doesn't offer the comfort and opulence we have come to expect of a Bond villain's hideout.[3] Blofeld's volcano lair makes no such compromises.

---

[2] Stanley Kubrick offered Adam a job on *Dr Strangelove* after watching *Dr No*, so the designer was unavailable for *From Russia with Love*.

[3] Raoul Silva's lair was based on the real-life abandoned island of Hashima, off the coast of Japan.

## Location, location, location

The idea for a volcanic home for a villain came to the 007 producers when they were scouting locations in Japan. Fleming's original plot for *You Only Live Twice* has Blofeld ensconced in a Japanese castle, surrounded by a poison garden designed to entice people in to kill themselves. It's a weird, sadistic start-up enterprise that he hopes to expand to locations all over the globe.[4] The filmmakers scrapped the idea completely and instead have Blofeld contracted by a mysterious foreign power to antagonise the US and USSR in the hope of starting a nuclear war. This third nation, heavily suggested to be China, can then emerge from the fallout as the dominant nation on earth. Filmed during the height of the space race, the plan is to kidnap orbiting US and Soviet spacecraft so that each side will blame the other.[5]

To accommodate this new scheme, Blofeld's lair needed an upgrade from a castle with a big garden to something with space to launch rockets and store the ones that have been stolen. Flying over various potential Japanese palaces and castles, the producers could find nothing that fitted the bill, until they spotted the peaks of a chain of extinct volcanoes. Ken Adam sketched an idea for a lair located inside one of the craters. Cubby Broccoli asked if he could build the set for a million dollars. Adam said yes.

---

[4] The garden was given to Safin in *No Time to Die*, along with the idea of exporting death globally (see chapter 025).

[5] There must have been easier ways to increase Cold War tensions in 1967, but this is the improbable, big-budget plan Blofeld and his paymasters went for.

The set was so big a special stage had to be constructed in the grounds of Pinewood Studios. This temporary structure needed more steel to hold it up than the London Hilton Hotel and came complete with a retractable roof over an opening big enough to allow a helicopter to land inside. At 45m (148ft) high, it could be seen from nearly 5km (3 miles) away, and the interior was so cavernous it took nearly every light, lamp and torch the studio had to light scenes that spanned 137m (450ft). The monorail actually worked, but the rocket landings were filmed in miniature and expertly blended into the final film.[6]

Building a pretend volcano lair was challenging enough, but a piece of cake compared to doing it for real. As locations go, volcanoes have a lot in their favour, at least from a villain's point of view. A solid, imposing structure, a readily available source of power and an inbuilt self-destruct mechanism. But there are also one or two downsides.

The surface of the planet we inhabit is just the cooled, hardened outer crust of the hot, flowing magma underneath. This hard, rocky layer is cracked into vast tectonic plates that are slowly shifted by the convection currents moving the magma underneath. Volcanoes are like vents along the cracks in this crust. These geological pressure valves allow magma to escape to the surface, sometimes slowly, sometimes spectacularly.

---

[6] The technical skills to build these sets is incredible. Technicians would turn down other jobs so they would be available for the next Bond film, not because it paid better, but because they wanted to work on them.

An erupting, or very active, volcano is an obvious non-starter for even the most villainous of villain's lairs. One very good reason for avoiding very active volcanoes, and their craters in particular, is lava. At an average temperature of 1,000°C (1,800°F), lava makes the crater inhospitably hot. At the other extreme is an extinct volcano, one that has never erupted in recorded human history. This might seem like a safer option, but volcanoes are unpredictable things. Fourpeaked Mountain in Alaska was classified as extinct as it had not erupted in at least 10,000 years, but then it decided to resume activity in 1995. There are also volcanoes that operate at activity levels between these two extremes. Many people around the world live in very close proximity to active volcanoes, though usually on the fertile outer slopes rather than the interior of the crater.

Blofeld must have carefully chosen a volcano with just the right balance of activity to take advantage of the geothermal energy but not risk an eruption through a rocket launch. But having found the perfect, mostly inactive volcano, he wouldn't be able to move in straight away. Volcanic craters are usually filled with debris from the last eruption that would have to be painstakingly removed to make way for the high-tech control room, monorail etc. On the positive side, heat transferred to the rocks above from the hot lava deep below could be used to generate electricity and Blofeld would never have to suffer a cold shower in his unusual home.

Excess heat and unpredictable eruptions aren't the only things to worry about if you are living inside a volcano. Lava has a lot of gases dissolved in it that can escape through fractures in rocks and diffuse through permeable soil. The sulphur in these gases readily reacts with moisture in the air to produce sulphuric acid that would quickly rust all that shiny steel and corrode the mechanism of the retractable roof. Metal surfaces and mechanical components would need to be heavily protected from corrosion. Using an alternative, corrosion-resistant metal like tungsten would cut down on maintenance and, having the highest melting point of all metals, it wouldn't melt in the heat of a rocket launch. It would be fiendishly expensive and difficult to work with, but money and inconvenience appear to be no barriers for Blofeld. None of this would help with the fact that many of these sulphurous gases seeping and belching out of volcanoes are also very toxic. The hordes of minions, scientists, henchman, cooks and cleaners needed to build and maintain a volcano lair would all be working in potentially hazardous conditions, not that this would bother Blofeld one bit.

As an aside, where do the villains recruit their staff? Is there a minion recruitment agency or does Blofeld just put a card in the post office window? And how does he command such loyalty? The sight of 150 ninjas abseiling from the roof and I'd be the first to throw down my weapon and point towards Blofeld, but these guys fight to the death. You'd need to offer more than a steady income and a free boiler suit to retain such an incentivised workforce. The benefits package must be phenomenal.

The working conditions Bond villain employees have to endure leave a lot to be desired. On top of the lax day-to-day health and safety, together with the risk of being attacked by Bond and teams of highly trained combatants, there is also a very good chance of your

workplace exploding – a worryingly common event in
007's world (see chapter 011). This is only compounded
when your workplace is a volcano. As mentioned,
volcanoes can remain dormant for a long time but there
is no guarantee they will stay that way. Ideally, Blofeld
wants to destroy his lair at a moment of his choosing: if
and when his hideout is discovered and comes under
attack.[7] Having proofed his hideout against regular
rocket launches, which are essentially controlled
explosions, it might take something fairly hefty to trigger
an eruption. A nuclear device should do it, and Bond
villains always seem to have at least one of those lying
about the place. Never mind the hundreds of employees
that would die in the process. Blofeld's plan may be
foiled, as it always is, but as long as he has provided
himself with a means of escaping before the eruption
engulfs his whole enterprise, he will survive to bother
Bond another day. Other Bond villains have not been so
meticulous in their planning and often go down with
the ship.

## Sleeping with the fishes

In the 1977 film *The Spy Who Loved Me*, Ken Adam tried
to outdo his set-building records. His design for Karl
Stromberg's supertanker, the *Liparus*, a ship whose bows
could open to engulf submarines and is so big it needed
its own maglev monorail to get around it, simply wouldn't
fit on any sound stage. So they built a new sound stage,

---

[7] The technical knowledge of how to land a rocket was lost under
the lava, and no one has managed to achieve this in real life until
very recently, when SpaceX developed reusable rockets.

the largest permanent one anywhere, complete with a water tank to fit three three-quarters-scaled subs side by side. Impressive though the design of the *Liparus* is, it isn't as ambitious as Atlantis, Stromberg's multistorey research lab-cum-lair, which can submerge under the waves when the world is all a bit too much for him to bear.[8]

Located in the Mediterranean, off the coast of Sardinia, Atlantis can apparently sink to a depth of anywhere up to 3,000m (9,800ft). Water is heavy, and the more of it there is above the roof of your lair, the greater the weight pressing down on it and the stronger the materials you will need to prevent your home from being crushed like a Coke can.[9] Including big, open, opulent rooms inside your undersea structure only makes it more difficult to hold up the ceiling. Stiffness is all-important, to stop your base flexing and folding in the currents or collapsing in the swell of a storm. Going deeper reduces the risk of storm damage, but even at the shallow end, the pressure the structures must withstand are huge. Those materials are also going to be under chemical attack from the salt water they are immersed in. Atlantis must be able to withstand a lot of chemical and physical abuse. Lots of steel and thick, reinforced glass is the order of the day, but the steel would have to be carefully protected with water-resistant coatings to stop it rusting.[10]

---

[8] In reality, Atlantis was just a model – quite a big model, but not big enough for Stromberg to fit inside, let alone anything else.

[9] For every 10m (33ft) further under the water, there is roughly an extra atmosphere of pressure weighing down on you.

[10] Again, tungsten may provide an expensive corrosion-resistant alternative to steel. Its extreme density would also work as ballast, to help sink all those big, airy rooms beneath the waves.

Undersea labs, living spaces and restaurants really have been built.[11] Some of these submersible structures operate as deep as 100m (330ft), but they are small and shallow compared to Stromberg's Atlantis. Submarines obviously operate at much greater depths, and carry large crews, but the spaces are cramped and their time under the waves is limited. Atlantis needs to be comfortable and self-sufficient because, if Stromberg succeeds in his plan to make the land uninhabitable, he is going to have to stay down there for a long time.[12]

Stromberg's hope is that the nuclear war he aims to start will result in a nuclear winter. The detonation of hundreds of nuclear weapons will not only kill and injure millions of people directly; the radioactive fallout will damage existing plant life, and the dust blown up into the Earth's atmosphere from the bombs and the resulting fires will reduce the sunlight reaching the surface and cool the planet. Crops will fail and millions more will starve.[13]

In 2014 Phil Pauley put forward plans for a Stromberg-style underwater habitat named Sub-Biosphere 2. With eight living pods and systems designed to keep the habitat completely self-sufficient in fresh air, water, food and electricity, the inhabitants can, in theory, ride out

---

[11] For example, the Aquarius lab off the Florida Keys can host six people at a depth of about 19m (62ft) and is used to train NASA astronauts.

[12] Definitely for months and probably for years.

[13] Some models predict parts of continental North America and Eurasia could experience a fall in temperature of 20°C (68°F), meaning it would be below freezing during the growing season, potentially leading to a 90 per cent reduction in crop yields.

any catastrophe that might be happening on the surface. The technology exists to build it, but this ambitious engineering project is still in the concept stage.

Under water might be the best place to sit out a nuclear winter, but Stromberg's beloved sealife would not be immune to the consequences. The water itself will offer some protection against radiation, and the deeper you go, the better. The seas will also take longer to cool than the land, so aquatic life would be likely to hold out best against the global changes happening on the planet. But everything will be affected. Ocean ecosystems rely on algae and other plant life that needs the sun for photosynthesis. Stromberg would be able to watch the results through those huge underwater windows in his lair.

Stromberg took an extreme approach to sheltering from the consequences of his actions, but, if your plan is to wreak havoc on the world, it is best to remove yourself from it completely. Hugo Drax's personal protection plan is the most ambitious of all Bond villains. His lair is in space.

### Out of this world
If the crater of a volcano or the bottom of the ocean is inhospitable to *human* life, space can't support *any* kind of life. There is not enough oxygen, warmth, food or anything really to sustain it.[14] You must take everything with you; there are no resources to tap into in space

---

[14] Bacteria and microscopic creatures like tardigrades can survive such an environment but they enter a kind of stasis, putting life on hold until more favourable conditions arrive.

other than a lot of free solar energy. Transporting all this life-supporting stuff off the planet is not easy. If building lairs inside volcanoes or under the sea are incredible feats of engineering, they are nothing compared to constructing a space station.

Drax's legitimate business is building shuttles, and he has several contracts to supply various nations with space vehicles. He can therefore pass off transporting massive amounts of equipment and components for his secret space lair as test flights. But it seems strange that no one would have noticed or asked any questions about these unusually large payloads.[15] While other Bond villains can cut a few corners by having a general disregard for the health and safety of their staff, Drax has a vested interest in the people who share his space hideout. He is relying on his band of specially chosen beautiful people to repopulate the Earth once he has killed everyone else (see chapter 006). Drax will need to take a lot of equipment with him to keep everyone in tip-top condition because space is not the healthiest place to be.[16]

Micro gravity can cause all sorts of problems for bones, blood pressure and muscle mass. Astronauts on the real-life International Space Station (ISS) have running machines and they exercise to offset these issues, but Drax has gone for a much more convoluted solution. Perhaps as Bond's most laid-back nemesis, he wanted to

---

[15] The ISS has taken the collaboration of five nations, 40 flights and 13 years to be completed, though upgrades and additions continue to this day. It took two years and eight flights just to make it habitable.

[16] Though certainly better than an Earth that Drax has bombarded with lethal nerve toxin.

avoid the treadmill. Instead of moving himself, he got his whole space station to move for him to create a kind of artificial gravity. By rotating his space station, those inside will feel as though they are being pushed outwards. This centrifugal force mimics the force of gravity. Though this system has been proposed for space vehicles, particularly those involved in long-term missions, none has yet been built.

Gravity, or lack of, is not the only potential detriment to health that Drax needs to concern himself with. All that free solar energy he can use to power his lair comes with a downside. Radiation risks increase the further you move away from the Earth's protective atmosphere. With no atmosphere to intercept the radiation streaming towards it from the sun and the rest of outer space, astronauts are at an increased risk of cancer caused by cosmic rays. Depending on how long Drax needs to stay in orbit until his nerve gas has done its work and the Earth is habitable by humans again, this may not be a problem. But if his time in space coincides with some solar flares – powerful eruptions from the sun's surface – he might need to construct a part of his space station that is better shielded as a temporary safe haven.[17]

Thicker walls would at least offer additional protection from another space hazard – debris. Though less of a problem in 1977, space, certainly the regions around the Earth, has become increasingly cluttered. Apart from all

---

[17] Astronauts on the ISS sometimes only get a few minutes' warning of a solar flare to move themselves to a part of the space station with thicker walls that can protect them from these sudden bursts of radiation.

the working satellites, there are quite a few that aren't working, as well as the detritus and debris from collisions and unwanted bits ejected from rockets as they deliberately break up during launch. All these objects become potential missiles as they fly around the Earth in their own orbits at fantastic speeds. Even tiny fragments of dust can inflict an awful lot of damage on a spacecraft because they are travelling at thousands of miles per hour. Thicker walls mean more protection but they also mean greater weight and more difficulties in transporting everything into space in the first place.

The logistics of Drax's operation, quite apart from developing a lethal toxin to kill all of humanity, are staggeringly complex. Yet none of the huge numbers of people that would need to be involved have leaked his secrets. When his completed station is finally revealed to the world it comes as a surprise to everyone. How is this possible? I know it's far away, but it's HUGE!

Cameras, telescopes, radar stations and radio receivers have been trained on Earth's skies since the late 1950s, spurred by the developing Soviet and US rocket programmes, the launch of Sputnik 1 in 1957, and the increasing threat of intercontinental ballistic missiles. And, while there is a lot of sky, it would be difficult to miss such a large and elaborate space station. The film explains this away with a jamming device that Bond and Goodhead disable to reveal the station's whereabouts.

Radio jamming, transmitting a strong signal to override or block another, has been used since the Second World War to prevent people listening to enemy broadcasts, or to disrupt communications during a battle. Drax could use this to hide communications between his station and arriving shuttles, but radio jamming won't hide the space station itself. So what about radar?

Radar jamming, like radio jamming, fills a radar detector with noise so that objects can't be located. The downside is that the radar station knows there is something there because their screens are filled with noise, they just can't pinpoint it. If Drax is using such a device, he will simply alert people to his space station's existence. Then, a few telescopes trained in the right general direction and a clear night, and anyone would be able to see it.

The much smaller, real-life ISS is visible to the naked eye from Earth, at least at night. Light from the sun reflects off the station so it can be seen as a bright white dot, brighter than any star, moving across the sky.[18] Perhaps Drax is using some kind of cloaking device that hides the space station from scientists on the ground. This idea, popularised by the TV series *Star Trek*, is the stuff of science fiction, though science fact has tried to catch up. Some materials can absorb one frequency of light and emit another to make them invisible to certain wavelengths. This is the idea behind stealth technology (see chapter 018). But hiding from all wavelengths is not currently possible. Stealth planes, boats and space stations might not show up clearly on radar but they're not invisible to the naked eye.

Suspension of disbelief is all part of the game of watching a Bond film or enjoying another 007 Fleming story. Ordinary criminals, dictators or megalomaniacs

---

[18] Check out the schedule here if you want to see it for yourself: https://spotthestation.nasa.gov/

make use of scientists and specially constructed labs or buildings to carry out their crimes. What sets a Bond villain apart from all the rest is locating those scientists, labs and lairs at the top of a mountain, on a remote island or frozen lake, and finishing them to a standard of opulence to which all egomaniacs would like to become accustomed.

# 006

# On Her Majesty's Secret Service and Blofeld's bioterrorism plot

Bond villains never lack ambition. Some love gold, some love power, some love destruction and chaos. The same could be said for many people, but what distinguishes a Bond villain from a common-or-garden variety ne'er-do-well is scale. The criminal masterminds in 007's world operate internationally. And, although these bad guys rarely target civilians, their grand plans mean mass casualties would be inevitable. Hundreds, if not thousands, would be killed should any of their space lasers or atomic weapons actually be used and there would be many more deaths should any of the targeted governments retaliate against these attacks.

A handful of these megalomaniacs have gone even further, deliberately planning human slaughter on a global scale. Ernst Stavro Blofeld, Hugo Drax, Karl Stromberg[1] and Lyutsifer Safin[2] seem to have a particularly low regard for human life, having deliberately planned the wholesale destruction of millions, even billions, of people. Drax's plan in *Moonraker* is to kill everyone with a plant poison and repopulate the Earth with his own specially chosen beautiful people. In *On*

---

[1] Stromberg, in *The Spy Who Loved Me*, tries to start a nuclear war (see chapter 005).

[2] Safin wants to kill lots of people with nanobots but let evolution influence the new world order (see chapter 025).

*Her Majesty's Secret Service*, Blofeld doesn't restrict his deadly campaign to humans. He has developed a biological weapon, a virus that could destroy everything, plants and animals alike.

Drax's and Blofeld's plans, like the villains that created them, may seem overblown and their methods far-fetched. But they are not as ridiculous as some of the schemes developed in the real world.[3] And if Drax's poison and Blofeld's virus really could do everything they claimed, then the effects would be far more terrifying than the films suggest.

## A bit of background on bioterrorism

Nature is not always our friend. There are many bacteria, viruses and other pathogens whose survival results in the death of other living things, including humans. Ever since there have been wars their outcomes have been affected by bacteria and viruses causing disease among the troops. In the past, dysentery and typhoid thrived in the squalid conditions troops endured while on campaign, and at times caused more casualties than the actual battles. Some humans have sought to use these diseases to their advantage, from contaminating wells with dead bodies, to fourteenth-century plague victims being catapulted into the besieged medieval city of Caffa in the Crimea to infect the citizens inside.

---

[3] Consider the tests that were carried out in 1950 from the US's biological research station Camp Detrick, and involved cluster bombs stuffed with turkey feathers dusted with a fungus to destroy cereal crops.

The twentieth century saw the introduction of more 'refined' weapons. The diseases and means of delivery became more targeted but no less awful. And, though biological weapons were heavily researched, they were rarely deployed, even in tests. At the time 007 was thwarting bioterrorism plots on screen and in the pages of Fleming's novels, the public would have been more concerned with the threat of nuclear weapons.

Biological agents have both advantages and disadvantages from the perspective of warfare and Bond villains. They have no taste or smell and can incubate in an animal or plant for many days undetected before symptoms appear. The organism, be it virus, bacteria or fungus, can reproduce and go on to infect more hosts. In theory you can do an awful lot of damage with relatively little material. Blofeld's choice of a biological weapon in *On Her Majesty's Secret Service* means he doesn't need to go into mass production. Carefully select just a few targets for initial infection and the biological agent will do the rest. This is also the potential disadvantage of biological weapons: the bugs don't always do what you want them to do. Bacteria and viruses cannot distinguish between friend and foe. Vaccines and treatments need to be available to protect your own side from infection.

## A diabolical biological plot

Fleming introduced Blofeld's biological plan in his 1963 novel *On Her Majesty's Secret Service*. He has his villain ensconced in a secret laboratory on top of a mountain, where he and his scientists are cultivating all manner of animal pathogens.[4] Also in his secret science lair is a group of beautiful young women, recruited under the pretext of treating their allergies. Blofeld and his team succeed in curing the women and send them back home with a gift. It's a canister of a bacteria, virus or fungus, and instructions to spray the contents on animals at agricultural shows. The infection will then be transported to farms when the show ends, and will spread.

To ensure he got his facts straight, Fleming employed Joan Saunders. She ran the Writers and Speakers Research Agency, whose members would provide answers to writers needing help with tricky factual details. Fleming requested information about 'germ warfare', including the best types of diseases or pests to deploy, and also where the major agricultural shows were held in the UK. 'I realise that this is all very fanciful stuff, but with the help of expert advice I think I can make it more or less stand up,' he wrote. The result was several pages of the novel devoted to a lot of concise, detailed and accurate information about biological agents, their effects and the problems they pose for detection. To bolster credibility, he also included plenty of references to real-life research and military plans for biological warfare, perhaps drawing on some insider knowledge from his contacts in the intelligence services.

---

[4] Blofeld's Alpine lair sounds a lot like Schloss Mittersill, an international sport and shooting club that was taken over by the Nazis in the Second World War. It became a centre for 'Asiatic studies'. When the Schloss's owners returned after the war, they found thousands of skulls from Tibet, India and China stacked on shelves built into every room.

In describing the potential scale of Blofeld's threat, an example is given from the Second World War: 'Around 1944 the Americans had a plan for destroying the whole of the Japanese rice crop by the use of aerial sprays. But, as I recall, Roosevelt vetoed the idea.' This was a real plan that was never carried out, though not quite for the reasons Fleming states. The Americans had calculated that the entire Japanese crop could be destroyed by releasing 20,000 tonnes of a chemical rather than biological weapon, codenamed LN8, over their rice fields. However, the Americans were also developing atomic weapons and these were completed before the required amounts of LN8 could be produced. So, in the hope of bringing the war to a speedier conclusion, two atomic bombs were dropped. Had Japan not surrendered at that point, the rice plan may have gone ahead the following year.

Roosevelt's decision not to attack the rice crop might have been down to practicalities, but he did consistently veto the use of biological weapons during the war. It shows that biological and chemical weapons are seen very differently – almost too awful to use, even in the context of a world war that saw millions killed, both combatants and civilians, with conventional weapons.

The novel also acknowledges British research into biological weapons but rather downplays how much went on: 'We were indirectly concerned in the fringes of the subject during the war.' In fact, the British poured a considerable amount of time and resources into biological weapons. The most well-known experiments, though not mentioned in the novel or the film of *On Her Majesty's Secret Service*, are those carried out using

anthrax in 1942 on the uninhabited Gruinard Island off
the west coast of Scotland.

Anthrax is an infection caused by spores of the bacteria
*Bacillus anthracis*, with a terrifying mortality rate. In
humans, around 20 per cent of skin infections are fatal
without treatment, and a staggering 80 per cent die if
the spores are inhaled, even *with* treatment. In the 1942
experiments, anthrax bombs were exploded close to a
group of tethered sheep. The sheep died from anthrax
infection just a few days later. It was clearly an effective
weapon, but a little too effective. The island remained
contaminated for decades. In 1981 large volumes of soil
were removed. In 1986 tonnes of diluted formaldehyde
were sprayed over the whole island and yet more soil
taken away. Then a flock of sheep was left there to see
what would happen. They remained healthy and in 1990,
48 years after the tests, the island was deemed safe.

Large-scale biological attacks can have long-term
consequences. In the novel the target is Britain, but in
the film Blofeld goes global. World domination isn't such
fun if you can never leave your mountaintop lair to
enjoy it. Blofeld would need to pick his pathogens
carefully.

**A few modifications**
The film adaptation of the novel, released in 1969, is
very faithful to the original story but with two important
changes. Instead of a range of biological agents, Blofeld
has developed one virus, the 'omega virus', which causes
'total infertility in plants and animals'. And the women
being treated at his clinic, his 'angels of death', are from
all over the world, not just the UK. These women have

been chosen because they have allergies to a common food item in their part of the world. After their successful treatment, the woman from China can happily eat rice, the girl from Ireland is seen gleefully tucking into potatoes etc. When they return to their homes they can infect a greater range of food staples, creating more damage over a wider area and the opportunity to extort money from more governments. Blofeld can hold the world to ransom by threatening devastation, crop by crop and livestock breed by livestock breed, until governments – in fear of widespread starvation, economic collapse and all-round chaos – capitulate.

To prove he isn't making idle threats, Blofeld informs Bond he has already tested his plan on a smaller scale. In the novel, Blofeld carried out a trial run with Fowl Pest, now known as Virulent Newcastle Disease, or VND, to distinguish it from other pestilences that can affect fowl.[5] Fleming chose VND deliberately to add topicality. In 1962, when Fleming would have been writing his novel, over 1 million turkeys and 10 million chickens were slaughtered to contain a real-life outbreak in the UK.

In the film, Blofeld claims responsibility for an outbreak of foot and mouth disease in the UK the previous year. It was again a deliberately topical reference. Less than two years before the film's release, the UK really had been hit by an outbreak of this highly infectious disease.

Foot and mouth is caused by a virus that produces blisters in the mouth and on the hooves of cloven-footed

---

[5] VND is very contagious and has a high mortality rate, but not the 100 per cent Fleming claimed.

animals.[6] Although it rarely kills, it makes the animal lame and unable to feed easily. The first reported case in the 1967 outbreak occurred on 21 October.[7] A farmer noticed one of his sows was lame. Four days later a vet confirmed the diagnosis of foot and mouth, by which time 17 pigs on the farm were infected. Once the disease had been introduced it was spread locally by the wind, birds, rodents and other animals. A ban on moving livestock was rapidly introduced and 400,000 animals were slaughtered over a period of eight months to bring the outbreak under control. The economic impact was huge because of the combined effects on the agricultural and tourist industries.

The release of foot and mouth disease at multiple sites across the world would be devastating, but Blofeld is even more ambitious: 'Not just a disease in a few herds, Mr Bond, or the loss of a single crop, but the destruction of a whole strain, forever, throughout an entire continent.' He also makes it clear that his omega virus can infect humans, his ultimate trump card, but he suspects he will not have to go that far. Bond is now in a race against time to warn everyone and stop the angels of death before they reach their targets. But can one virus really do so much harm?

It is true that most viruses can infect more than one species but they are usually within the same class of organism. For example, some strains of influenza can

---

[6] Foot and mouth can infect humans but the symptoms are very mild.

[7] It is believed to have originated in legally imported Argentinian frozen lamb.

infect birds and humans, but they are both animals. Viruses must get inside their host's cells in order to reproduce, but plant and animal cells are very different. Humans are almost never vulnerable to viruses that infect plants, and vice versa.

If Blofeld's scientists have successfully engineered a virus that can infiltrate both plant and animal cells to cause an infection, this presents another problem. How will he be able to target one species at a time? If he selects chickens as his first target, how will he stop the chicken from infecting every other living thing it comes into contact with? Even a chicken farm needs humans to run it. Something tells me Blofeld might not have thought this one through. I'm not saying that it's a bad plan. I'm saying that one virus to kill them all is a really, *really* bad plan.

If his specially made omega virus really is as virulent as Blofeld describes, it could result in a widespread, uncontrolled disaster. He might shoot himself in the foot as the virus spreads beyond his initial targets, regardless of whether his ransom demands are met. His scientists must have developed antidotes, treatments or vaccinations against the omega virus if only to protect themselves and Blofeld from exposure to it. Auctioning this knowledge to the highest bidder might be the easiest way to extort money from different governments. But, even if treatments are already known and available, it takes time to mass-produce them. Hospitals soon become overwhelmed, even if the infection is incapacitating rather than fatal. Preventative measures, such as mass vaccination programmes, take months to prepare and deploy. It is hardly surprising that most governments have shied away from the use of biological weapons but continue research into them for defensive measures.

Blofeld's plan could be a disaster for everyone, including Blofeld. It would be more plausible if the filmmakers had stuck to the original multi-pathogen plan outlined in Fleming's novel. It should go without saying that mass extermination is something to be avoided. Unless, of course, your name is Hugo Drax.

## Criminal chemical plans

Ten years after *On Her Majesty's Secret Service* hit the big screen, James Bond foils another biology-inspired evil plan. This time the villain, Hugo Drax, not satisfied with large-scale slow slaughter through starvation, plans to wipe out the entire human race as quickly as possible. To achieve his goal, Drax has weaponised the fictional rare orchid, *Orchidae nigra*.

According to Q-branch, the orchid contains a chemical that can cause infertility in humans, but Drax's scientists have modified it to make it deadly. We get a clear indication of the lethality of this toxin early in the film when a phial full of it breaks inside a laboratory. Bond watches through a window into the hermetically sealed room where two scientists clutch at their throats and collapse on the floor. Two rats in a cage in the same room are completely unaffected. Drax's poison can kill a person in seconds, but no other animal or plant will be harmed by it.

The chemical structure of Drax's enhanced orchid poison is displayed on a screen, and at first glance it is an impressive attempt at showing some real chemistry. Looking closer there are a few mistakes but, to the filmmakers' credit, there are parts of the compound that look as though they genuinely have the potential to do some harm. In the middle of the molecule are three fused hexagonal rings that, if you squint a bit, are like a steroid structure. Hanging off one end of the molecule is something not unlike an organophosphorus compound, the kind of molecular structure that forms the basis of the VX nerve agent discussed in chapter 004.

His toxin may have been derived from a plant, but because Drax is planning to use a specific compound rather than a whole organism, he is about to wage chemical rather than biological warfare. The differences between the two are important. Chemicals can't reproduce themselves the way bacteria or other pathogens can. Every human will have to be directly exposed to the lethal compound to suffer its effects. To do that, Drax has built 100 spheres, each filled with enough toxin to kill 100 million people. Each globe would presumably explode high above the ground so that the toxin will rain down over the largest area possible. These toxin–filled globes will be launched from space. It means he can target sites all over the world while keeping himself and his team of beautiful people out of harm's way.

Although much of the film *Moonraker* is outlandish, the idea of mass extermination using chemical weapons is horribly credible. The gas attacks carried out in the First World War and the use of cyanide in the Nazi gas chambers in the Second World War are stark reminders of what humans can inflict on one another knowing full well the horrific effects of the chemicals they are using. Drax's grand plan for mass extermination has obvious parallels with the Holocaust, but the method he has chosen is closer to the gas attacks of the First World War. Between April 1915 and November 1918, tens of thousands of tons of chlorine, phosgene and mustard gas were released from cylinders to drift towards enemy trenches. Thousands died and thousands more were left permanently disabled by their effects, but toxic gas was not as efficient a weapon as many had hoped – the gases could collect in pockets in some areas but be dispersed by the wind in others.

Research was carried out into many other toxic substances, ones that were more deadly or could be distributed more effectively because they were liquid or solids. One substance, ricin, received a lot of attention because it was thought it might be better suited to large-scale, open-air operations.

Ricin is a plant-based chemical like Drax's orchid poison.[8] In fact, it is a protein produced by *Ricinus communis*, a popular ornamental plant and the commercial source of castor oil. The protein is made up of two chains of amino acids, A and B, linked by a single bond. Inside the human body, ricin can latch onto the surface of cells and open a channel to the interior. Once inside the cell, the two chains break apart. Released from its partner, chain A can interact with ribosomes, molecular machines within all cells that build the proteins and enzymes our bodies need to function properly. These ribosomes are essential for our day-to-day health, growth and repair, and ricin stops them from working, and not

---

[8] It also has spy associations from its use to kill Georgi Markov on Waterloo Bridge in London (see chapter 004).

just a single ribosome. One chain A can bring 1,500 ribosomes to a standstill every minute. It means a staggeringly small amount of ricin can do a lot of damage. When cells cannot replicate or repair themselves, everyday functions break down and the cell dies. As more cells die and there is no way of replacing them, organs stop working, leading to death in a matter of days.

Castor oil plants are easy to grow, the ricin is easy to extract and there are no antidotes. Left out in the open it can also be broken down by the UV light in sunlight over a few days, unlike the anthrax discussed earlier. It means that had Drax chosen ricin as his weapon, once he had succeeded in obliterating humankind from the surface of the planet he wouldn't have to wait long before he could repopulate it with his handpicked beautiful people. Ricin is exactly the kind of thing you would expect a Bond villain to be producing in bulk to take over the world, but there are considerable practical difficulties to overcome.

Being a protein, ricin is easily and permanently denatured by heat, rendering it useless – think what happens to a steak or an egg when you cook it. During the Second World War, the Americans found ways of producing ricin on a huge scale and grinding it using specially chilled milling machines to prevent the heat of friction from denaturing it. Conventional bombs can't be used for dispersal as they generate too much heat, so the British developed bomblets that could spread clouds of ricin dust effectively. Ricin could be produced in bulk and distributed widely, but there were still problems.

The personal risk ricin poses depends on the route taken to get inside the body. Getting ricin into the bloodstream, via cuts or a deliberate injection, is the most effective way of delivering it to cells so it can do its damage. In this scenario, less than 1mg can be fatal, but it is difficult to execute on a large scale. Inhaling ricin particles into the lungs is also very dangerous and might occur in the immediate aftermath of a ricin bomb, or if particles distributed by the bomb are disturbed. If you were to eat it, your digestive system would treat it like any other protein and start to break it down into its component parts, rendering it ineffective. At least a hundred times more ricin is required to kill via this route.[9] Huge amounts of ricin would have to be added to food or water supplies to pose a risk, and simply boiling the water or cooking the food would remove that risk. But, in the case of a bomb, the likeliest point of exposure is the skin, and ricin can't be absorbed into the body that way.

The Americans decided ricin had too many problems and felt phosgene gas was more suited to their wartime needs. Though the French also started researching ricin, they quickly dropped their plans, considering it too great a risk until an antidote could be found. Without an antidote, the only practical way of using ricin would be to disperse it remotely and keep yourself at a safe distance for several days. Dropping bombs from a secret space station in orbit would be the ideal scenario but was beyond Second World War technology or budgets. In the fantasy world of 007 there no such restrictions, and, fortunately, there are also secret agents like James Bond to save the day.

---

[9] Still only about 100 mg. This is in no way a recommendation to eat it. Please don't eat ricin or feed it to anybody else.

# *Diamonds Are Forever* and diamonds

Classy, desirable, seemingly indestructible and often caught up in villains' evil plans, diamonds and 007 have a lot in common. These iconic stones are scattered all over the James Bond franchise: in briefcases paying for drugs or military hardware, adorning the villains' pet cat or iguana, or embedded in the face of their henchman. The seventh film of the series is positively saturated with them. They're in the title, Shirley Bassey sings their praises in another classic belter of a theme song, their smuggling is the key thread in the plot and they are used to make the villain's secret weapon.

Ian Fleming also got a lot of mileage out of diamonds, not just in *Diamonds Are Forever*. He penned several newspaper columns and a non-fiction book centred on the jewels and their attraction to criminal enterprises in particular. Beauty, glamour, thefts and smuggling rings would seem to go with the territory as far as diamonds are concerned, but a giant space laser? Let's take a look at the gems that have been idolised in books, films and songs. What makes them so special and could they really be weaponised to hold the world to ransom?

**Diamonds are a spy novelist's best friend**
Fleming's love affair with diamonds started in 1954. After spotting 'a diamond is forever' advert in a copy of *Vogue* he picked up at an airport, diamonds became the subject

of several columns he wrote for the *Sunday Times*, full of his characteristic detail: 'I suppose De Beers think of diamonds simply in terms of carbon burned into the form of crystals in which each atom is tetrahedrally linked to four others at distances of 1.54 Angstrom units'.

He was right about the arrangement of the atoms, and the distance between them. The tetrahedral configuration means the carbon atoms in diamonds are arranged in the most closely packed and interconnected way possible. Each diamond is a single incompressible crystal of carbon, scoring the maximum 10 on the Mohs scale of mineral hardness, as Fleming also noted later in his article. He was also correct when he wrote that diamonds burn to produce carbon dioxide gas ('perish the thought', he adds). Carbon is carbon. The structure of diamonds certainly makes them more difficult to burn than other forms of carbon, but burn they will, and they can ignite at temperatures as low as 690°C (1,270°F), well within the range of an ordinary house fire.

They also have some interesting optical properties. Because of the way the carbon atoms are arranged, pure diamonds are completely transparent to visible light and therefore perfectly clear and colourless. As light moves through the lattice of atoms it is diffracted into a rainbow of colours, giving diamonds a unique sparkle. Diamonds' properties are impressive, certainly, but not extraordinary. So how then can they command such eye-watering prices?

Carbon has very little intrinsic value. It is common enough on this planet but very little of it has been arranged into diamonds. The intense conditions needed to compress the atoms into the dense, three-dimensional lattice that is diamond, only occur naturally deep within the Earth's crust. Natural, gem-quality diamonds were formed billions of years ago in the intense heat and pressure found more than 150km (93 miles) below the Earth's surface. Millions of years ago, kimberlitic eruptions forced streams of gaseous magma up from the depths of our planet through pipe-like structures to the surface. Some of these kimberlite pipes drilled their way through diamond-bearing rock, bringing those that survived the dramatic changes in temperature and pressure to the surface. Of the roughly 6,000 kimberlite pipes that are known around the world, only a few dozen contain enough diamonds to be worth mining.

Rarity will increase the value of many things. Having a use will also give a material worth, but diamonds fulfil no role so well or exclusively that we couldn't survive without them. The hardness of diamonds makes them a useful addition to grinding and cutting tools, so small diamonds, that contain flaws or are an unappealing colour, are reserved for these industrial applications.[1] Large, clear diamonds, however, have absolutely no practical use, but people are prepared to pay huge amounts of money for them. Their value is entirely down to their desirability. Diamonds have a status above and beyond other gemstones, for reasons partly cultural, partly fashionable and partly down to marketing.

Diamonds' toughness and brilliance has ensured they have maintained their position as high-status objects. As affluence increased in the twentieth century, more and

---

[1] Coloured and cloudy diamonds are caused by impurities and imperfections.

more people could afford to buy the jewels. De Beers worried that if lots of diamonds were suddenly released onto the market they would be devalued. Their response was ingenious. In 1948 a marketing campaign was launched with the phrase 'a diamond is forever'.[2] A diamond engagement ring became the best and only way to demonstrate your everlasting love. Grand gestures and the promise of eternity came with a reassuringly hefty price tag. Demand for diamonds rocketed and the price could be kept profitably high.

The increasing number of diamonds to be found in private homes sparked a corresponding surge in the theft of these valuable jewels. In 1907, the *London Daily News* ran a story about Arthur Edward Young who had pleaded guilty to several charges of burglary in the Streatham area of London. Young was noted for his climbing abilities and had acquired the nickname the 'cat burglar'. The name, and the crime, caught on.[3] Jewellery was easy to carry and could be broken up into its components for resale. Removed from their settings, the gems were essentially untraceable. Diamonds were, of course, the most desirable target thanks to their incredible value.

The temptation to steal diamonds at source, from the diamond mines, was even greater. Uncut diamonds don't look particularly special and are small enough to secrete

---

[2] Even though they're not. Diamonds are very hard but can be broken and, even if they escape fires, all diamonds are slowly disappearing into the more stable form of carbon – graphite. It takes thousands of years though, so no need to worry about your family jewels.

[3] The phrase 'cat burglar' had been used before 1907 but in reference to people who stole cats.

about your person. It seemed easy enough for those working in diamond mines to take a few uncut diamonds for themselves. Surely no one would miss one stone among so many, and the theft of a single diamond the size of a marble could set you up for life. The problem was that the diamond companies did miss those stones and organised tight security to keep hold of them. Access to diamonds was easy, they were everywhere; the trouble was getting them out of the mining compound and into the hands of a buyer. Elaborate ruses were developed, bribes were handed over to security and medical staff who carried out checks, as well as couriers, pilots and boat owners who could help smuggle the stones to willing buyers.

The combination of glamour, danger and crime was too good for Fleming to confine to a few newspaper columns. He decided that diamonds and diamond smuggling would be the focus of one of his spy novels.

## Diamonds Are Forever

The plot of Ian Fleming's fourth James Bond novel centres on a fairly straightforward diamond smuggling ring run by the Spangled Mob, headed by Jack and Seraffimo Spang. Their network of criminals siphons off diamonds mined in Africa, through Europe and into the US. Bond takes the place of a diamond smuggler to infiltrate the gang and put a stop to the whole racket.

In addition to what he had learned writing his newspaper columns, Fleming incorporated other ideas and gathered yet more information about diamonds and smuggling before settling down to write his 007 adventure. One idea came from Charles Frazer-Smith, who used hollowed-out golf balls to conceal messages sent to prisoners of war (see chapter 002). Fleming swapped the messages for diamonds and had Bond smuggle the stolen gems past US customs in his golf bag.[4]

Fleming also made enquiries directly with De Beers, who allowed him to watch the sorting and cutting of diamonds in their London branch. He also talked to Sir Percy Sillitoe, a former head of MI5 who had been employed by De Beers to run an anti-diamond smuggling operation.[5] Further verisimilitude was added with accurate descriptions of locations around the US, gathered from Fleming's first-hand experience.

Not much of Fleming's novel survived the adaptation to film. The basic idea of diamond smuggling is still there, as are a handful of the minor characters and the main locations. This framework has been padded out with every tried and tested Bond trope and traditional ingredient until it's bursting at the seams.[6] Not satisfied with one Bond woman with a pun in place of a name, Tiffany Case, she is joined by Plenty O'Toole. Fleming's Spangled Mob is replaced by not just one Blofeld, but lots of Blofelds.[7] Everyone had loved the laser in *Goldfinger*, so why not bring it back, but make it bigger

---

[4] In the film he uses the dead body of the smuggler he is impersonating.

[5] Perhaps as a way of thank you, Sillitoe gets a name check in the final novel.

[6] An overcompensation for the perceived disappointment of *On Her Majesty's Secret Service*.

[7] The villain has taken advantage of plastic surgery, which was growing in popularity at the time, to make copies of himself.

and more powerful? Better still, the laser could be launched into orbit. Why settle for a paltry few million from diamond smuggling when billions could be extorted from the world's governments by threatening to destroy their submarines and missile silos from space?

The boring reason why not, is because giant space lasers are not a thing. At least not yet. The film tells us that Blofeld's diamond-studded laser satellite has been designed and built under the supervision of an expert in light refraction, and diamonds certainly have remarkable refraction properties. We are told that the laser is generated through the first diamond, and because there are lots of them stuck all over the satellite's disk, the combination will produce one very powerful beam. In 1971 diamond space lasers were pure science fiction, but science fact has started to catch up.

The first diamond laser was built in 2008, a prototype to show how the unique properties of the material could benefit laser systems. Diamonds have the advantage of being transparent across a very wide range of wavelengths of light and could therefore be used to tune lasers to emit different colours. For example, yellow/orange beams that are particularly useful in eye surgery but very difficult to produce with conventional lasers, could be more easily generated using diamonds. Diamonds also have a greater ability to amplify light, known as optical gain, than any other material, increasing the potential power of the laser.

Increasing the power also increases the heat generated, but diamonds have excellent thermal properties too and can conduct the heat away more efficiently than other laser mediums. It is the reason why diamonds are sometimes referred to as ice because they quickly conduct heat away from the hand when you touch them, making them feel very cold. It makes the method chosen for smuggling diamonds in *The Living Daylights* particularly appropriate. In this Bond film the clear, colourless gems are mixed in with the ice keeping a donor heart cold. The customs officer, disgusted by the sight of the heart, waves the package through without taking too close a look.[8]

The potential for diamond lasers is huge but any wannabe Blofelds out there can't just prise a diamond out of the family jewels and start building weapons to take over the world. The diamond crystals used in these lasers need to be of exceptional quality to maximise their properties. Natural diamonds are full of defects that make them unsuitable for laser applications. Artificial diamonds are what you need.

The first reliable method for growing synthetic diamonds was invented in 1954 by Tracy Hall. He designed a press that could apply 100,000 atmospheres of pressure to squeeze powdered carbon, heated to 1,600°C (2,912°F), into a tiny crystal of diamond.[9] The majority of the synthetic diamonds made this way are used for industrial purposes because they are small and dark in colour, but new techniques have changed that. Chemical vapour deposition is a method where very

---

[8] Obviously customs officers wouldn't normally be opening up boxes containing donated organs as this would expose them to potential contamination, particularly on a dusty airfield, but it is fiction.

[9] Hall's employers, GE, made millions out of the process. Hall was rewarded with a $10 savings bond.

high-purity carbon is turned into a gas and allowed to condense onto a tiny seed crystal in a controlled way. The technique can be used to build up bespoke, optically pure diamonds layer by layer.

More diamond lasers have been built since the 2008 prototype and they have been improved with each new iteration. Diamond lasers now compete with traditional lasers in terms of efficiency and are opening up all sorts of new applications, some of them in space. It has been suggested that diamond lasers could be used for communication systems between satellites as well as for tracking and clearing space debris. However, Blofeld's space laser is still in the realms of science fiction.

With sufficient energy, lasers can cut through metal (see chapter 003), but a staggering amount of power would be needed to create the kind of space laser Blofeld's scientists build. Quite apart from the technical difficulties of launching the device into space and operating it remotely, it would need to generate a beam with enough energy to cut through many kilometres of Earth's atmosphere and still be able to heat up huge metal structures like missiles when the beam made it to ground level. Not to mention making its way through several metres of sea water on top of that to explode a submarine.

Space lasers are out of the question, for the time being anyway. When diamonds have reappeared in later 007 films they have been more credibly placed in the hands of criminals using them as a form of currency. Fleming would probably have approved as he developed a particular interest in diamond smuggling.

## Stranger than fiction

Following the publication of *Diamonds Are Forever*, Fleming was approached to write a factual piece for the *Sunday Times* about the subject of this novel. His interviews with John Collard, a former MI5 agent who had been employed by De Beers to put a stop to the smuggling, were written up and serialised as 'The Diamond Smugglers'.[10] Fleming invented a suitably Bondian pseudonym for Collard, John Blaize, and gave him the romantic but entirely fictional job title of 'diamond spy'.

Collard's stories covered all manner of ingenious methods for stealing and smuggling diamonds, and how they were stopped. There were tales of miners flinging diamonds over security fences with catapults, tying them to homing pigeons, slipping them into their socks, wedging them into the treads of their shoes or car tyres, even swallowing them.

Diamonds in socks or stomachs are easy to spot with X-rays. The stones are transparent to the X-rays used in hospitals and airports, showing up as black dots.[11] Some workers tried to fool the X-ray machines that scanned their luggage every time they left on holiday by hiding their stolen diamonds inside lead containers shaped to look like everyday objects. Others, just as in Fleming's fictional tale, tried to bribe staff to look the other way.

Many of Collard's stories were uncannily like something out of a James Bond book. He told Fleming

---

[10] It was also published as a book with the same title.

[11] Workers couldn't be X-rayed too often without giving them a lethal dose of radiation. The trick was to make them believe they were being X-rayed even when they weren't.

about a crime boss he named Monsieur Diamont. He claimed Diamont was the biggest crook in Europe, perhaps the world, and operated beyond the reach of the law. He promised Fleming that if he were to show up in Diamont's neighbourhood having published some of the stories Collard had told about him, 'he'd have you bumped off'. The description of Diamont, a German by birth, as 'a big, hard chunk of a man with about ten million in the bank', could easily apply to Fleming's fictional Bond villain Auric Goldfinger.

Despite the incredible stories, helped along by Fleming's engaging writing style, 'The Diamond Smugglers' was not quite the enthralling read it promised. Fleming complained, 'It was a good story until all the possible libel was cut out.' His literary association with diamonds was at an end, but Bond and the big screen weren't done with them yet.

## Diamonds in the rough

The 1983 Bond film *Octopussy* centres on a circus, run by the glamorous and enterprising woman who gives the film its name. The circus is being used as a front to smuggle jewellery. The jewellery smuggling is in turn being used as a cover to transport a nuclear device into Germany, a scheme the rogue Russian General Orlov hopes will start a war. The bit about the nuclear device was invented, but the circus and the jewellery smuggling were taken from a real-life case almost as ludicrous as the one portrayed on screen.

In 1982, a scandal broke involving members of The Moscow State Circus. One of the performers, known only as Boris the Gypsy, an extravagant character fond of ten-gallon hats and cowboy boots, was arrested along with Anatoly A. Kolevatov.[12] Kolevatov oversaw a number of circus, ballet and ice skating companies. He was the one who chose which acts went abroad on tour. Apparently the idea was that the circus would perform abroad for a fee, but that fee would be paid in jewels. When Kolevatov's apartment was searched, over a million roubles in diamonds were discovered along with other jewels and foreign currency. Despite being one of the world's largest producers of diamonds, Russia also had a very healthy black market in the stones and Russians were seemingly happy to pay well over the odds for these already very valuable objects.

Russians are far from alone in going to great lengths to obtain diamonds. Sierra Leone has been both blessed and cursed with natural diamond deposits. In past decades groups of rebels have fought to control the country's diamond fields. The battles between different factions have destroyed lives and left many dead. The gems these groups have sold to fund their activities have become known as blood diamonds or conflict diamonds. Legitimate diamond traders are keen to distance themselves from these activities and have sought many ways to reassure their customers that their diamonds are not blood diamonds.

Though every diamond is made of carbon, all natural diamonds contain small amounts of impurities. Up to

---

[12] Others were also caught up in the intrigue, including General Zotov, head of the bureau responsible for issuing emigration permits, and the daughter of Leonard Brezhnev, Gloria, who had once been married to a circus performer.

about 1 per cent of a diamond can be comprised of other elements. The identity and ratio of these trace elements are characteristic of where they came from. All that is needed is a chemical survey of the diamonds from each source to form a library of data for comparison. This is how, in *Die Another Day*, Bond uncovers Gustav Graves' evil plan.

In homage to *Diamonds Are Forever*, diamonds take on a similar central role in this 2002 Bond film. There is even a giant space weapon, although this one is funded by diamonds rather than powered by them (see chapter 003). The diamonds in question are said to have come from a newly discovered source in Iceland and have been marked with the Gustav Graves logo to prove their authenticity, an idea taken from real life.

De Beers uses a logo etched by laser onto their diamonds. As the company sources all its diamonds from its own mines it can be assumed that any stones etched with their logo are not conflict diamonds. More recently, Tiffany & Co. have announced that all their new diamonds will have a microscopic serial number etched into them with a laser. The unique number can be used to find out a whole host of information about the history of the diamond, from where it was mined to where it was cut. In the film, the logo is a flat-out lie and cover for Graves' illegal activities. Analysis of the diamond's chemical composition reveals they originated in Sierra Leone and are therefore likely to be conflict diamonds.[13]

---

[13] The film shows Bond and his contact Raoul inspecting the diamonds through magnifying glasses, but you wouldn't be able to see the elements this way. You would need quite sophisticated lab equipment to properly analyse the trace elements.

*Die Another Day* has diamonds not only being smuggled and used by the bad guys to fund terrifying space weapons, but they also adorn the face of one of the henchmen, Zao. In the pre-title sequence Bond blows up an attaché case full of diamonds that were part of an illegal arms deal. The blast causes some of the stones to be thrown out of the case and into Zao's face. When Bond meets Zao several months later the diamonds are still embedded in his skin.

Diamonds are chemically and biologically inert, meaning that should an attaché case full of them explode in your face, one of the few things you wouldn't have to worry about is an allergic reaction. The diamonds would be completely ignored by the skin around them. Skin has a very high rate of renewal; it needs to, as it is the outer covering protecting us from the wear and tear of everyday life. Old skin cells slough off to be replaced by new skin cells underneath. As the skin healed and renewed itself, the diamonds in Zao's face should have fallen out. Maybe, like thousands of others before him, he chooses to keep the diamonds because he likes them. Despite the efforts of scientists, real and fictional, to find a practical use for diamonds, their main purpose remains as personal adornment and status symbols – perfectly Bondian attributes.

# 008

# *Live and Let Die* and the crocodile run

From parrots to piranhas, and cats to crocodiles, the world of James Bond contains a surprisingly large and varied menagerie. Bond villains may have no compunction about killing humans, but they are big animal lovers. Lethal or unusual pets go hand in hand with the 007 bad guys, but not all the animals are malevolent.

Franz Sanchez's affectionate iguana in *Licence to Kill* is simply decorative. The horses in Zorin's stud farm in *A View to a Kill* are part of the cover for his evil plan.[1] Max the parrot in *For Your Eyes Only* is positively helpful, telling Bond and Melina where to find the bad guy Kristatos. Blofeld's fluffy white cat may have a vicious streak a mile wide but its capacity to do harm is limited to what it can get its claws into.[2] The huge Dobermanns owned by Hugo Drax in *Moonraker* are a different story. They are well trained and obedient, and chase down and kill Corrine Dufour only on command. These dogs are closer to what you might expect of a Bond villain's pets: a nice addition to make their secret lair feel more homely but with the capacity to be weaponised when necessary.

---

[1] But Zorin also uses biological trickery to enhance their race performance so he can make even more money on the side.
[2] The stunt cat used in *You Only Live Twice* was so traumatised by the sound of Blofeld's gun going off it never worked again. The cat wrangler was furious.

It's not just dogs that have been used to guard the bad guy's lair or laboratory. Though you might think shark pools would come as standard in any Bond villain's hideout, they are not all so conventional. Should you outlive your usefulness, at least in the eyes of a bad guy, you might be dispatched by scorpion, piranha or snake. Bond himself has had to fend off attacks from tarantulas, sharks and tigers. One of the most extravagant protective pets are Dr Kananga's crocodiles, used to guard his drug lab in *Live and Let Die*. The crocodile farm provides the set-up for one of the 007 films' most memorable stunts – Bond running, in crocodile shoes, across the backs of five live crocodiles.

There are many exotic creatures stalking and slithering about in the 007 franchise but they can be broadly grouped into two main categories based on the type of threat they pose. First, there are chemical threats from animals that attack or defend themselves by use of venoms. And then there are those that pose a physical threat with their sharp teeth and/or claws.

You could also include Blofeld's cat in a category all of its own, as it deserves to be. Blofeld has been portrayed on screen by five different actors, some scarred, some unscarred, some with a full head of hair and some bald, but all with a white cat.[3] The effect of this cat is psychological but powerful; its appearance usually signals something bad is about to happen, but the animal itself is of little direct threat. And though this feline warrants many words dedicated to its iconic status, this chapter

---

[3] Although he isn't allowed to take the cat with him when he is imprisoned in Belmarsh in *No Time to Die*.

focuses on the more lethal pets put to use in Bond's adventures.

## Animal attraction

Bond has some strong wildlife connections thanks to his creator Ian Fleming. The secret agent's name was appropriated from the author of *Birds of the West Indies*,[4] a book Fleming always kept on hand when in Jamaica. Fleming also enjoyed swimming over the local coral reefs to watch the fish, but all sorts of wildlife interested him. Barracuda, sharks and other marine life as well as birds and land animals feature prominently in his stories.

*Dr No* features a bird sanctuary and lake borrowed from Inagua (see chapter 001), as well as a giant squid, spiders, a centipede and crabs. When it came to adapting the story for the screen, a few changes were made. Giant squid are incredibly reclusive animals and there was no chance of obtaining a live one for filming. Any prop giant squid would look ridiculous, so the enormous cephalopod was unceremoniously cut. Crabs were easier to obtain and having them crawling over Honey Ryder's body, as they do in the book, would make an arresting visual. The problem was, the crabs weren't really interested in acting, or doing anything much, and some died under the hot studio lights. The scene was adapted to have Ryder threatened with drowning in the slowly

---

[4] In a letter to the wife of the real James Bond, Fleming offered 'unlimited use of the name Ian Fleming for any purpose he may think fit. Perhaps one day he will discover some particularly horrible species of bird which he would like to christen in an insulting fashion that might be a way of getting his own back.'

advancing tide, and the crew got to take home a lot of crab for dinner.

The huge centipede placed in Bond's bed to bite and kill him should have been much easier to replicate.[5] In the book, Quarrel explains to Bond that Jamaica is home to some nasty centipedes that can kill people. And while it is true that some centipedes can give very nasty and very painful bites, there is only one confirmed death by centipede, a seven-year-old Filipino girl who was bitten on the head by a *Scolopendra* centipede.[6] The filmmakers decided to use a tarantula instead. Perhaps they were hoping to ramp up the tension by using an animal that many people already have a phobia of, and it would look impressive on screen. Sean Connery, however, refused to have the spider anywhere near his bare skin and a sheet of glass was placed between him and the arachnid. His stunt double was used for the shots of the spider actually crawling on Bond's arm.

Spiders are certainly responsible for more human deaths around the world than centipedes, but tarantulas, though they do have a venomous bite, are more likely to defend themselves by flicking irritating bristly hairs towards their attacker. An encounter with a tarantula can therefore be unpleasant but not lethal. Nevertheless, rumours of deadly South American tarantulas persist, perhaps due to confusion between 'wandering' spiders or 'banana' spiders such as those in the genus *Phoneutria*.

---

[5] An idea almost certainly stolen from Sax Rhomer's *The Insidious Dr Fu Manchu*.

[6] In 2014 in Venezuela a four-year-old died after reportedly being bitten by a giant centipede.

These spiders are large, hairy and responsible for many trips to hospital, particularly in Brazil. *Phoneutria* often wander into homes, shoes and bunches of bananas seeking shelter and can give a very painful bite if disturbed. Their venom contains neurotoxins that stimulate nerves, resulting in vomiting, oedema, tachycardia and a lot of pain, but treatment is possible and the prognosis is usually very good even without resorting to antivenoms.[7]

Dr No may have been happy to use a spider to attack Bond, but the filmmakers were hardly likely to risk using an animal that could send their star to hospital. Where venomous creatures have formed part of a Bond plot, either fake animals have been made by the props department, such as the blue-ringed octopus in *Octopussy*,[8] or non-lethal varieties have been cast in the role.

**Vicious venoms**
Snakes, like spiders, are animals likely to produce a visceral reaction in cinema audiences. The problem was they had a similar effect on most of the cast and crew working on the set of *Live and Let Die*. The person responsible for continuity refused to be present whenever a snake was involved in a scene (which was often) and had to have events described to her afterwards. Not everyone could avoid the snakes so easily. The unfortunate Dennis Edwards, playing Baines in an uncredited role,

---

[7] The exception is young children, where there is a small but significant risk that the effects of the venom could kill.
[8] See chapter 002.

was tied between two posts and menaced with a snake as part of a voodoo ritual scene. When the snake, a non-venomous emerald tree boa, was held close to the actor, he collapsed, not because he had received a lethal bite as suggested by the plot but because he had fainted from fear. Geoffrey Holder, who played Baron Samedi, was also terrified of snakes and only flung himself into a coffin full of non-venomous varieties because he didn't want to look scared in front of some royal visitors who were on set that day.[9]

Fear of snakes is somewhat justified. Snakes are the biggest animal threat to human life, being responsible for an average 50,000 deaths annually worldwide.[10] The majority of the snake deaths will be as a result of the venom they inject through their fangs. The toxic contents of snake venom vary between species, and even individuals of the same species, but they are complex mixtures of compounds that can affect nerve and heart function and destroy cells. There are some nasty combinations in the snake world that can kill humans very effectively, but not as fast as seen in *Live and Let Die*. To make them even more terrifying, snakes have other methods of killing, as shown in a later Bond film.

In a scene in the 1979 film *Moonraker*, a bridge over the villain's ornamental pond gives way, with Bond on top of it. He falls into the watery home of the villain's pet python but manages to kill it after a dramatic struggle.

---

[9] The other snake that slithers into Bond's bathroom in *Live and Let Die* is a non-venomous speckled king snake.

[10] Mosquitoes are indirectly responsible for the most human deaths: around a million people every year, due to the diseases they transmit.

The big fake snake has the markings of a reticulated python, which are native to Southeast Asia. But this snake-infested pond is situated in South America, where they have their own native big snakes, and ones with a much deadlier reputation. Anacondas have suffered many slurs against their nature thanks to many films that have portrayed them as man-eaters. They are certainly known to eat very large animals, but the only recorded case of a snake eating a human involved a reticulated python. *Moonraker's* baddie went the extra mile, but others make do with what is to hand.

In *Diamonds Are Forever* Mr Wint and Mr Kidd take advantage of local wildlife to dispatch a troublesome dentist. The psychopathic pair have been sent to Sierra Leone to cover the tracks of Blofeld's diamond smuggling racket. They find a big black scorpion under a bush and extol the virtues of 'mother nature's finest killers'.

There are approximately 1,500 different species of scorpion, over 30 of which are considered a very real threat to humans. These 30-plus species deliver around 1.2 million stings to humans each year, leading to 3,200 deaths, a fatality rate of 0.27 per cent. Wint and Kidd have perhaps overestimated the animal's lethality. There are many factors that determine if a scorpion sting will be fatal. First, the amount of venom delivered. Dry stings deliver no venom at all; the scorpion is merely going through the motions as a warning against a perceived attack. Second, size. Adults with bigger body mass are less likely to be killed than children. And finally, how rapidly the individual can access treatment.

In the film, the scorpion is dropped down the shirt of Dr Tynan, who has been siphoning off diamonds from the mine where he is employed. The result is agonised screaming before the dentist collapses. The agonised screaming is understandable: a scorpion trapped inside a shirt may well be provoked to sting, and inject a considerable amount of venom with that sting. Scorpion venom produces a very painful reaction thanks to the collection of nerve poisons it contains. But the instant death? Scorpion venom can spread rapidly through the body, symptoms can appear very quickly and a person's health can deteriorate suddenly, but we are talking about minutes and hours rather than seconds.

If Dr Tynan passed out from the pain of the sting and was simply left for dead, his chances of survival in a remote location without any medical help would not be good. There are antivenoms available – of particular importance when treating children – but treatment of symptoms can also improve the prognosis. The intense pain but relatively low mortality, and availability of treatments, may be why scorpion stings were chosen to torture Bond in *Die Another Day*, though his captors presumably aren't too concerned about his eventual fate.

Sharing your workplace with deadly creatures makes some sense when your business is torturing secret agents for information. Less explicable is the decision to keep lethal animals captive within a commercial property where you would presumably want your human clientèle to feel as safe and welcome as possible. In *Skyfall*, a pair of komodo dragons are kept in a pit in a Macau casino.

For many years it was assumed these huge lizards killed their prey (and occasionally harmed humans) through their bite causing blood loss, combined with the effects of toxic bacteria in their mouths. However, research has shown the mouths of these animals don't harbour any particularly nasty or unusual bacteria, but they do have venom glands. This venom contains a mix of proteins including those associated with haemorrhage and shock, consistent with reports of prey animals becoming immobile or docile after being bitten. Komodo dragons don't have a particularly strong bite but they do have sharp serrated teeth that can open up large wounds as they yank their prey backwards, as happens to the casino security guard about to shoot Bond. The venom is secreted from openings between their serrated teeth and into the wound, but how much this venom contributes to the death of their victims is still not fully established. The eventual fate of the guard is unknown as Bond doesn't stick around to find out, escaping the dragon pit by jumping on the back of another of the creatures slowly advancing towards him.

Komodo dragons are a kind of crossover between the venomous and biting animals so popular with Bond villains. The James Bond books and films are littered with creatures with very sharp teeth and healthy appetites for humans, but do these animals really deserve their fearsome reputation?

## All the better to bite you with, my dear

In a film franchise that leans quite heavily on generalised character traits and conventions, it can be no surprise that it isn't just the humans who suffer from stereotyping. The filmmakers regularly take advantage of some animals that are simply 'known' to be dangerous, even though it isn't always the case.

*You Only Live Twice* has Blofeld living in his magnificently appointed private volcano. The expanse of desk from which he makes his plans for world domination is protected by a piranha-filled pond. Visitors to this inner sanctum must cross a bridge over the pond, conveniently fitted with a trapdoor. How worried should his SPECTRE agents be after receiving a summons to his office?

Stories of shoals of piranhas attacking humans and stripping their flesh to the bone in seconds, as claimed in the film, are persistent but wildly exaggerated. These stories may well have been started by no less a person than Theodore Roosevelt who, while on a tour of Brazil, witnessed a demonstration of the piranha's biting ability. A cow was pushed into a pool full of piranhas and rapidly skeletonised. But, to get the desired dramatic effect, the piranhas had been collected in a shallow pond formed by damming off part of the river and starved for several days.

These freshwater omnivorous fish can certainly administer a painful bite – the black piranha (*Serrasalmus rhombeus*) has the most powerful bite of all bony fish – but they rarely cause death.[11] Children are more vulnerable to attack as splashing attracts the fish, and their smaller bodies mean piranhas can do comparatively more damage. Hungry and stressed piranhas are also more likely to attack. If Blofeld is stressing and starving his fish then SPECTRE agent Helga Brand may well be in trouble when she falls through the trapdoor. But, having had a whole adult human to dine off, the fish

---

[11] A bite force of 320N has been recorded.

were probably sated by the time Hans the henchman falls into the waters later in the film.

Piranhas have an impressive bite but the champion fish chompers are sharks. Almost any Bond film featuring an underwater sequence shows a shark or two milling about in the background, but Bond villains also have a nasty habit of keeping these animals in their private pools. Sharks are apex predators that have successfully survived on this planet in one form or another through two major extinction events and over 420 million years. They are formidable creatures, but they are a poor choice for a pet, even if you are a Bond villain.

Sharks aren't easy to keep in captivity. Many species range over vast distances and dive to incredible depths to hunt out their prey. A small pool and the occasional henchman to feed on is just not the same. Despite their man-eating reputation, sharks are not natural aggressors towards humans. Of the dozens of shark attacks reported every year, many are provoked, and virtually all bites from sharks are exploratory or defensive. If they really did have a taste for human flesh, sharks would make very short work of the thousands of people that bathe in their natural habitats around the world. The result would be a death rate from shark bites much, much higher than the current average of four per year. The experience the 007 crew had when filming with sharks in *Thunderball* illustrates the point.

In this film, the villain, Emilio Largo, has decided to fill one of the pools in his beachside home with the entirely fictional species of 'grotto' sharks. Largo describes them as the 'most dangerous species of shark' but that dubious honour is normally given to bull sharks because, as the only species of shark that can survive in fresh and salt water, they are simply more likely to be found in the same place as humans.

The real sharks put in the pool for the film were pretty listless. Kevin McClory, one of the producers, delighted in jumping in the water and poking them with a stick to provoke them.[12] To reassure Sean Connery when he had to get in the pool, he was told a clear plastic barrier was in place between him and the sharks. What they didn't tell him was that they didn't have enough plastic, and there was a 1.2m (4ft) gap the sharks could, and did, swim through. The look of shock on Connery's face as a shark swims towards him required no acting whatsoever. There were some close encounters, but none of these sharks attacked any of the many humans that shared the pool with them during filming.

Plenty of future Bond villains would feature shark pools, and plenty of henchmen would meet their demise in their depths. But death by shark is not inevitable, not even in a Bond film. Felix Leiter's encounter with sharks in *Licence to Kill* is lifted directly from Fleming's novel *Live and Let Die*. In the book, Fleming goes into considerable detail about sharks and their behaviour and describes the snuffling, grunting sound of a shark feeding from personal experience,[13] but he stops short of describing the actual attack on Leiter. The filmmakers

---

[12] That was until one of the sharks turned and bit his stick in two, after which he was less enthusiastic about getting so close to them.
[13] He and a friend went shark fishing off the Jamaica coast with the carcass of a dead donkey for bait.

decided to show it and in surprisingly graphic detail. There is lots of churning water, screams and clouds of red to tell you everything you need to know about the horrors going on in Krest's shark tank, but Felix survives the ordeal.

The human body can sustain a surprising amount of damage, as long as it is to the periphery. Arms and legs can be lost without killing someone. Superficial wounds to the head and torso can also be survived. Damage to major organs is more immediately life-threatening but they are protected to some extent by our bony ribcage and skull. But, needless to say, the greater the damage or the more disassembled a person is, the less likely they are to survive.[14] With rows of razor-sharp teeth, sharks can leave severe injuries even with only tentative bites.

Great whites, the shark everyone recognises from its man-eating role in films like *Jaws*, can grow to over 6m (20ft) long and have around 300 serrated teeth in its jaws, which can bite down with a force that may exceed 1.8 tonnes.[15] Only one Bond henchman would be a match for such an impressive animal – the appropriately named Jaws. In *The Spy Who Loved Me*, Jaws is dropped into Stromberg's shark pool. To deliver a twist on the well-worn trope, the henchman bites the shark rather than the other way around.

Sharks usually just ignore humans. If you want to swim with these magnificent animals the water around

---

[14] Of course, a lot depends on having access to suitable and speedy medical intervention before someone bleeds to death from their injuries.

[15] Based on computer modelling.

the shark cage often must be baited in order to attract them. It is the kind of strategy employed by Kristatos in *For Your Eyes Only*. In another scene taken from the novel *Live and Let Die*, Bond is tied to Melina, his love interest in the film, and the pair are dragged behind a boat over the sharp coral to rip into their flesh. The appearance of the unusual bodies in the water and the blood and flesh from their wounds are supposed to attract the sharks that Kristatos hopes will finish them off. It's certainly more likely to work than the few drops of blood employed by Dr Kananga in the film version of *Live and Let Die*.

Dr Kananga, like so many Bond villains before and after, has a shark pool in his lair. Bond and his love interest, this time Solitaire, are about to be dunked into the shark's home, but not before his arm is cut so blood will drip into the water and entice them. Sharks do have a phenomenal sense of smell, but their ability to detect minute amounts of blood in gallons of water and over vast distances is greatly exaggerated. It is believed that most of their prey detection is a combination of following smell trails, left in an animal's wake as it swims, visual inspection and detecting the electric fields generated by all living things, particularly when they are stressed or thrashing around. We will never know if Kananga's shark plan would have worked, as Bond escapes before the sharks attack. But some of the other creatures he uses to guard his criminal operations pose a more credible danger.

To keep his drug lab safe from unwanted attention, Kananga has located it in the middle of a crocodile farm. The Nile crocodile has a bite strength comparable to a great white shark but is responsible for far more human deaths.[16] These animals may appear sluggish and indifferent, but it is a myth that they will not attack unless provoked. Despite their large size, they are agile and very fast over short distances, even on land. They ambush their prey, clamping down on the nearest part of the body, be it arm, leg, head or torso. Most attacks are in defence of territory or young but, unlike sharks, large crocodiles do see humans as potential food.

When the 007 producers were scouting locations for *Live and Let Die*, they spotted a warning sign by a crocodile farm: 'Trespassers will be eaten'. The owner of the farm, Ross Kananga, had inherited it from his father after one of the livestock ate him. Unperturbed, Kananga junior supplemented his income from farming these animals by wrestling them. It was ideal James Bond material. The producers used the location, the owner's name, his crocodiles and his expertise.[17]

Wanting to make the most of this wonderful find, the filmmakers wrote a special scene, trapping Bond on a tiny island in the middle of a pond full of hungry crocodiles and only a few pieces of raw chicken to defend himself with. Tee-Hee, the head henchman, watches from the shore and offers tips on how Bond can fend off an attack. All Bond needs to do is jam a pencil in the tiny

---

[16] Records of crocodile and alligator attacks are not nearly so complete or detailed as those kept for shark attacks, but these animals are believed to be responsible for around a thousand human deaths every year. Nevertheless, attacks are very rare.

[17] Before filming, Kananga starved the crocodiles for 14 weeks to keep them lively because, as Roger Moore noted in his film diary, 'a well-fed croc is a quiet, sleepy croc'. Crocodiles can go for many months without eating anything.

hole behind the crocodile's eye. Bond notes he has no pencil. The second method is 'twice as easy' Tee-Hee says, laughing, 'you just reach into their mouth and pull out all their teeth.'[18] Bond is left standing in his crocodile shoes with fewer and fewer pieces of chicken and more and more crocodiles advancing towards him. It's a brilliant scenario and even the filmmakers couldn't figure out how their hero would escape. It was Ross Kananga who came up with the simple but brilliant solution of running over the crocodiles' backs to make it to land.

Ross Kananga was duly dressed in Roger Moore's costume and sent out to the island. Five crocodiles were held in place in the water by tying their legs together, but with their heads and tails still free to move. On the first run, he slipped and fell in the water. Moore suggested putting running cleats on the shoes to give him a better grip. Three more runs were filmed and each time the crocodiles were learning. On the fifth, the one that can be seen in the film, the crocodiles are anticipating his arrival and whip their tails and snap their jaws as he approaches. Kananga was bitten on the heel and the filmmakers decided they'd got enough. It is what the Bond films do best: collaborative efforts and expertise to create seemingly impossible scenarios and memorable stunts.

---

[18] Crocodiles and alligators are best left alone and at a distance. However, should you be attacked, the advice is to fight back in the hope they will give up and let go. Gauging the eyes is thought to be your best bet. If you can reach them.

# 009

# *The Man with the Golden Gun* and the golden gun

In a franchise full of innuendo and cutting quips, there is one thing Bond never jokes about – his weapon. He takes comfort in holding it while he sleeps and gets tetchy when others criticise its performance. Few can match Bond's prowess with his gun. Then along came Francisco Scaramanga to challenge 007 as the world's top shot.

*The Man with the Golden Gun* makes some uncomfortable comparisons between Bond and the villains he finds himself up against. Bond and Scaramanga both kill people for a living, but one is earning a million per shot (near enough $34 million in today's money) and the other is on a meagre government salary.[1] One trains in a cleverly designed fun house on his private tropical island against skilled opposition; the other goes to a nondescript rifle range in the basement of the MI6 building and shoots paper targets. One has a luxurious bespoke weapon and ammunition; the other has a standard-issue mass-produced gun and bullets. And Bond is the one everyone is supposed to envy?

---

[1] In the 1955 novel *Moonraker* Bond says he earns £1,500 a year as a Principle Officer in the Civil Service. This is equivalent to roughly £40,000 a year today. It might not seem like much but there is a lot to be said for a steady income and a good pension.

Guns are the tools of 007's trade and almost everyone he encounters has a weapon concealed about his or her person. It is unsurprising that firearms of all sorts are such a big part of the books and films. They feature on the jacket covers, and publicity photos invariably show Bond holding a gun. The incredible range of firearms that exists, as well as the ammunition they can hold, presented a golden opportunity for Fleming to indulge in the detail and brand-name dropping that characterises his stories.

However, I, and I suspect quite a few other people watching Bond films, would struggle to tell a Walther PPK from a water pistol.[2] I do know it is the weapon of choice for a double-0 agent, and I also know Bond prefers his beloved Beretta. Most of us get our information about guns from TV and cinema. And, though I don't want to shatter any illusions for you, sometimes they exaggerate a little. The details of make and model may pass me by, but the scientific principles that make all guns work are the same. This chapter is all about Bond's equipment, and the hardware everyone else is packing in 007's world.

## Tools of the trade

Ian Fleming had an interest in guns (he owned several himself) but seems to have appreciated them more for their craftsmanship and symbolism rather than the finer

---

[2] For instance, I did not know that in Sean Connery's publicity shots for *From Russia with Love* he is holding a Walther air pistol rather than a PPK. The barrel may be long, but it's narrow, and the gun is far too low-powered for Bond.

details. He was certainly no expert. For his first novel he wrote to a gunsmith, Robert Churchill, to review the names of the firearms he had mentioned in the book. The only one he'd got right was the .38 Colt Police Positive because he owned one (a gift from General Donovan during the war 'for special services').[3] The other three were all wrong. Even the personal weapon he had given to Bond, a Beretta, was spelled incorrectly and had the wrong calibre.

With spelling and calibre corrected, Bond got along well enough with his Beretta for five novels. Then Geoffrey Boothroyd, a gun enthusiast based in Glasgow,[4] wrote a fan letter to Fleming. While Boothroyd declared a love of the 007 adventures, he was distressed by Bond's 'deplorable taste in firearms'. In his comprehensive letter he dismissed Bond's Beretta as a 'ladies' gun' and suggested a Smith & Wesson Airweight to be a firearm more suited to a secret agent. The Airweight was light enough to carry about Bond's person and, being hammerless, it wouldn't catch on his clothing when he needed to draw it. He also recommended something heavier for Bond's car, a .357 Smith & Wesson Magnum, which would be more effective for long-range shooting. He went on to suggest suitable holsters and weapons Bond's opponents might be likely to carry on them (German Lugers, Polish Radons or Russian Tokarevs). From bullet calibre to barrel length, his recommendations

---

[3] Fleming also owned a 12-bore shotgun and a Browning .25, issued during his wartime service. He took the Browning with him on his holidays to Jamaica.

[4] He was a member of several gun clubs and owned 45 weapons, which is very unusual for someone living in the UK.

were full of the detail Fleming loved and the pair entered
into a correspondence.

In Fleming's next novel, *Dr No*, the Beretta is taken
away from Bond and replaced with a Walther PPK, a
scene recreated in the film. Fleming also took the
opportunity to pay his thanks to Boothroyd by giving
his name to the character of the Armourer, the man who
supplied Bond's weaponry. Major Boothroyd, more often
called Q, became a fixture of the film franchise. But
despite his enthusiasm, Fleming managed to muddle
things up. The Airweight was added to Bond's car instead
of the Magnum, and the wrong holster was slung under
Bond's arm.[5] Any letters of complaint Fleming cheerfully
passed on to Boothroyd.

When it came to the filming of *Dr No*, Boothroyd
was brought in to check everything was in order. Though
few in the audience would notice if the wrong character
was holding the wrong weapon, in some situations it was
vital for the plot. For Bond, detailed knowledge of guns
is essential. In *Dr No*, he is faced with a man pointing a
gun towards him who has instructions to kill Bond. But
Bond has counted the bullets already fired from the gun
and notes 'that's a Smith & Wesson, and you've had your
six', before shooting him. Knowing that his opponent's
gun is empty, Bond effectively kills an unarmed man in
cold blood. To make matters worse, he fires another
bullet into the dead man's back. This is unsporting,
unethical, cruel and definitely not cricket. It caused
outrage with the censors.

---

[5] The holster he chose, a Berns–Martin triple-draw, was for
revolvers and the Walther would have fallen out.

For the most part, we as the audience don't need to know the finer points of firearms. Any detail that might be relevant will be explained to us. For example, on board Goldfinger's private jet, Pussy Galore points a Smith & Wesson .45 at Bond. He responds: 'the bullet will pass through me and the fuselage like a blowtorch through butter. The cabin will depressurise and we'll both be sucked into outer space.' We know Bond is always right, but are the filmmakers?

## Whether it's a Walther or a Smith & Wesson

All firearms work on the same basic principle: rapidly expanding gases from a chemical explosion drive a projectile down a barrel. However, every aspect from the chemicals used to create the explosion, to the means of firing it, the projectile it pushes and the barrel it travels down, can be changed and adapted to meet different criteria for different weapons. This means there is a fantastic array of firearms and an even more bewildering range of ammunition to fire from them. It also means that the basic components of a gun can be adapted to fit inside other objects, like ski poles (*The Spy Who Loved Me*), cigarettes (*You Only Live Twice*)[6] and walking canes (*The World Is Not Enough*). Conventional weapons, however, dominate the 007 franchise. Sometimes guns so big and complicated you assume they must be the product of the fertile imagination of the props department are real weapons. The Walther WA 2000 rifle

---

[6] The British Special Operations Executive (SOE) apparently really did develop a .22-calibre one-shot gun that was concealed in a cigarette.

handed to Bond to carry out an assassination at the start of *The Living Daylights* has so many extras and bits bolted to it, it's almost a disappointment it only fires bullets. As M might say, 'is that all it does?' This rifle is apparently accurate up to 1,000m (3,280ft). Bond needs to hit a target across the street. Some might say he was overcompensating.

Though the choice in weapons is great, they can be grouped into a few broad categories. There are automatic weapons that keep firing bullets until either the trigger is released or the magazine runs out of ammunition. These types of guns tend to be quite large, and in 007 films they are most likely to be found in the hands of anonymous henchpeople or military personnel involved in the big battle towards the end of the picture. The main protagonists usually have smaller handguns, but they get more close-ups and so feature more prominently.

To the English, all handguns are pistols.[7] Some might be single-shot, like Zukovsky's walking-cane gun; others are revolving pistols (so called because a rotating cylinder moves a new round into the firing position), such as Goldfinger's gold-plated Colt. The rest are self-loading pistols, like Bond's Walther PPK, or the Walther .32 ACP (Automatic Colt Pistol). These handguns are often referred to as 'semi-automatics'.

The first pull on the trigger of a self-loading pistol raises the hammer and releases it into the firing pin, which slams into the primer in the bullet. The primer ignites and the bullet is propelled down the barrel by the

---

[7] Americans are more relaxed in their terminology, calling handguns revolvers and pistols.

expanding gases. These gases are also used to push the slide to the back of the gun, ejecting the empty cartridge case, inserting a new bullet into the firing chamber and cocking the hammer, ready to fire again. Subsequent pulls on the trigger are a lot easier because much of the work has already been done.

Any character could mark themselves out by picking a rare or specialist gun, but the bulk of the audience is unlikely to notice and appreciate such sophistication. If you want to show off your weapon, as any Bond villain naturally would, you need something that is going to catch everyone's eye. The answer is always gold. In the novel, Goldfinger uses a gold-plated Colt Model 1908 Vest Pocket.[8] In Fleming's novel, *The Man with the Golden Gun*, Francisco Scaramanga carries a single action Colt .45, again gold-plated.[9] If they were made of pure gold they would be unfireable, as gold is a soft metal and would deform with the first shot.

In the film adaptation, Scaramanga's gun has been given an upgrade, along with Fleming's plot. As a sideline to his assassination work, Scaramanga is attempting to supplement his income by selling the secrets of the solex agitator, which can harness solar power, a theme that tapped into the oil crisis happening at the time of filming. Scaramanga still has a gold-plated Colt, but he uses it only to open champagne bottles. For assassinations he has a bespoke, gold-plated weapon that

---

[8] In the film, Goldfinger fires a Colt Official Police, but still gold-plated, of course.

[9] These characters were created, after all, by a man who typed his novels on a gold-plated typewriter. Goldfinger wasn't the only one who loved gold.

comes in kit form. The components of the gun are disguised as everyday objects – a cigarette case, a lighter and a pen – which he assembles for each operation.[10] Of course, such a specialist weapon must also use specialist bullets.

### Firing blanks

While guns are important, their effectiveness also depends on the ammunition they fire. All sorts of ammunitions are produced, but most have the same basic components. There is a cartridge, a cylinder usually made of brass, closed at one end to contain the primer and propellent, and with a projectile, or bullet, clamped into the other end. The projectiles can, and have, been made of almost every conceivable material, depending on the application.

Magnesium can be used to make flares; explosives can be added to ignite on impact (see the exploding bullets in *You Only Live Twice*); poisons, like the cyanide darts in *Moonraker*, can be added to make them even more deadly; or compressed paper can be packed into a cartridge for the blanks used in filming the many gun battles in Bond movies. But by far the most common material for a bullet is lead. It is readily available and soft enough to easily form into the desired shape. Sometimes, though, lead can be too soft, so it is hardened by alloying it with other metals, such as antimony or tin. The bullet can also

---

[10] There were actually two golden guns used in the film, made by jeweller J. Rose, one fully made up and another that could be assembled and disassembled. Christopher Lee practised assembling the gun in front of the TV at night.

be jacketed or coated with a thin layer of a harder metal, often nickel or steel, to prevent it from deforming.

Usually, the projectile is designed to inflict damage on a target, and there are two main factors that determine how much damage can be caused: speed and size. The bigger the surface area pushing through tissue, bone, blood vessels and organs, the more widespread the destruction. But bullets don't necessarily enter the body point first and continue to drill a neat hole through the interior.

Bullets have a pointed shape to improve their aerodynamics – travelling further than the spherical bullets of the past. The rifling inside gun barrels, the spiralling grooves cut into the interior that you can see in the gun-barrel sequence in every Bond film, impart spin on the bullet as it travels along its length. This spin gives the bullet greater stability in the air and improves the accuracy of the shot. However, there is still a wobble, or wag, as it travels through the air, especially towards the limit of its range, meaning a bullet can hit at any angle, even side-on. As a bullet moves through the body it tumbles, putting strain on its structure and increasing the chances it will fragment and create more projectiles to tear more paths through the body.

To create maximum surface area, some bullets are designed to change shape when they hit their target. Lead is soft enough that it can easily deform on impact.[11]

---

[11] Dumdum bullets were originally standard rifle bullets that had the front of the metal jacket trimmed back to expose the lead core. These bullets were outlawed by the 1899 Hague Convention for military use, but not for civilians or police forces.

This deformation can be increased by not fully jacketing the bullet and leaving a cavity at its tip. Hollow-point bullets are designed to expand their diameter, up to 200 per cent, on impact, increasing the wounding effect.

Scaramanga would never contaminate his golden gun with anything as base as lead. The pure 24-carat gold bullets he uses in the novel are harder than their lead equivalents, but still pretty soft. In the film, chemical analysis reveals his bullets are slightly impure 23-carat gold with traces of nickel to make them even harder.[12] If the idea is to inflict maximum damage with a single shot, hardened gold seems like a poor choice. It's an expensive, and potentially risky, affectation, but expensive and risky affectations are a Bond villain's trademark.[13]

The other way to inflict as much damage on your target as possible is to increase the speed of the bullet. Any bullet travelling through the body pushes the surrounding tissue out of the way. Because tissue is elastic, a temporary cavity is created, much bigger than the diameter of the projectile, that rebounds once the bullet has passed through. High-velocity bullets create a massive shock wave and huge temporary cavities that stretch tissue beyond its elastic limit, causing extensive damage.

In the film, Scaramanga's gun is a single-shot. He needs to make sure that the bullet he fires will kill, or at least incapacitate, his target quicker than it takes him to

---

[12] Patrice's bullets in *Skyfall*, made from depleted uranium, a metal so hard it is usually used to make armour-piercing ammunition, is even less likely to deform (see chapter 023).

[13] There really are bullets known as 'golden bullets' but these are ordinary lead bullets with a copper coating.

reload. High-velocity gold bullets will undoubtedly cause a lot of damage, but the amount of damage does not always equate to stopping your opponent. It all depends on where the damage occurs. Being shot rarely causes someone to collapse instantly as seen in the case of so many henchmen, unless there is damage to brain function. If Scaramanga is half as good as he claims to be, targeting the head shouldn't be a problem, and his hardened bullet would easily penetrate the hard bone of the skull and pulp the brain. However, accelerating bullets up to speeds that can maximise damage presents another problem.

How fast a bullet can travel has a lot to do with the propellant. One gram of nitrocellulose, the traditional base of bullet propellants, can produce over a litre of gas which, confined inside a gun, causes an explosive force pushing the bullet down the barrel. The hot gases emerge from the muzzle behind the bullet and, released from the confines of the barrel, expand faster than the speed of sound. The result is a loud noise and a bright flash that can alert people to the whereabouts of the gun. Drawing attention to your location is the last thing you want when in the process of assassinating someone. This doesn't seem to worry Scaramanga, but Bond is often seen screwing a silencer into the end of his weapon.

## Less bang for your buck

Silencers are more properly called sound suppressors because they cannot silence a gun completely, only reduce or disguise the sound of it firing. A gun fired with a sound suppressor sounds like a muffled crack or a car door being slammed, not the soft 'phut' we hear in cinemas. Real sound suppressors are also much larger than the cigar-sized tube seen in 007 films. They work by slowing and disrupting the expansion of the gases emerging from the muzzle through a series of baffles combined with an expansion chamber.[14] These devices can be fitted to the end of any gun but are of limited use when screwed into the barrel of a pistol. While it will reduce noise from the gases escaping the muzzle, there is a gap between the barrel and the cylinder at the other end where gases can escape and create noise.

The noise of a gun comes not just from the expanding gases but also the bullet. If it travels faster than the speed of sound there will be a sonic boom, or crack. The only way to eliminate this noise is to reduce the speed of the bullet by using subsonic ammunition, venting the gases from the barrel, or using wipes, to slow the progress of the bullet.

Geoffrey Boothroyd, in his letters to Fleming, was particularly scathing about the use of silencers. He thought so few people knew what a gun sounded like when it was fired, it was pointless going to the trouble of using a sound suppressor. However, Fleming insisted they were a necessary tool for Bond's job. The real secret service was certainly interested in developing better silencers, as we saw in chapter 002. According to Fleming, British Intelligence had developed a silencer for use on a Sten gun that was so effective 'all one could hear was the clicking of the machinery'. He might not have been making this up. The Welrod was a single-shot gun developed during the war that had a cushioned firing

---

[14] Other designs are available.

pin and a sound suppressor built into the barrel. The loudest noise coming from the gun was the mechanism, which couldn't be heard at all from a few metres away.

With so many guns being fired, Bond, and almost everyone else in his world, must have become familiar with the sound of a sound-suppressed gun. These barrel extensions would be rather redundant in disguising what you were up to. If the use of silencers is an effort to preserve the user's hearing, it is strange that health and safety considerations haven't been extended to the bullets flying around almost everywhere Bond goes. Neither he nor anybody else takes any real precautions about getting shot beyond hiding behind something or ducking. We almost never see a bulletproof vest. Only two people have worn them in 60 years and 25 films.[15] One is General Pushkin in *The Living Daylights*, and the other is the sensible CIA agent Pam Bouvier in *Licence to Kill*.[16]

## Using protection
In *The Living Daylights*, General Pushkin collaborates with Bond to fake his death. He wears small bags filled with fake blood that burst from the impact of Bond's bullets. The bullets are stopped short of releasing real blood by the bulletproof vest underneath. Bulletproof vests, better termed bullet-resistant, are made from a range of materials but the most well known is Kevlar.

---

[15] Brad Whitaker's faux military uniform in *The Living Daylights*, made from bulletproof cloth, is too ridiculous to count.

[16] In *Die Another Day*, Bond holds up a Kevlar vest to shield himself from General Moon's bullets as they hurtle along on a hovercraft, but he has never spoiled the cut of his suit by actually wearing one.

Developed by Du Pont in the 1960s, Kevlar is a polymer fibre that can be woven into sheets of different thickness and densities depending on the application. For soft body armour, sheets of Kevlar are assembled into 'ballistic panels' and sewn into a vest. The woven, layered fibres first deform the bullet, increasing its surface area and transferring as much energy as possible from the bullet to the surrounding fibres. The impact will certainly be felt by the wearer, but the bullet can be stopped from penetrating the body. For example, Pam Bouvier in *Licence to Kill* is knocked over, apparently dead, from a gunshot to her back but soon recovers.

In *The Living Daylights*, whether Pushkin is knocked out by the impact of Bond's bullets or faking doesn't matter; he survives. But it's quite a high-risk strategy. Not only must Bond be an excellent shot to hit the vest,[17] but the bullets he fires must deform enough to spread the energy as much as possible. A bullet that doesn't deform can push its way through the fibres to enter the body. This is why Bond uses steel-tipped bullets against a KGB sniper at the start of the film because they are known to wear body armour. Extra protection can be given by inserting metal plates into the vest, but it becomes heavier and more cumbersome to wear.[18] It would have been much safer for Pushkin not to rely on Bond. It would have been as simple to rig the blood bags with squibs, miniature explosives that would break the

---

[17] Not such a worry; this is James Bond we are talking about.

[18] The metal cigarette case that stops a bullet in *From Russia with Love* doesn't count.

bags at minimal risk to Pushkin. This is how gunshot special effects are created for films.

Bond never wears body armour, but his car gets a lot more protection – perhaps a reflection of how much Q values his work over his colleague. The main body of a car can be strengthened using thicker, harder metals or composite materials, but the most vulnerable parts are the windows. Glass is brittle. A window will shatter if a bullet is fired at it, but the bullet will scarcely be slowed in its progress or deflected from its path towards its target. Glass can be toughened by laminating two sheets together with a layer of plastic in the middle. In this case, when a bullet hits the first layer, some of the energy of the bullet will go into breaking the glass and flattening the tip of the bullet. The glass will break, but the plastic layer will act like glue and hold on to the shards. The bullet will continue to push through the material, but with a bigger surface area that will disperse more energy to the material it is travelling through, and it will be slowed. The more layers of glass and plastic, the greater the resistance. A single layer of laminate will stop a 9mm-calibre pistol bullet, but Bond would need a lot more layers to protect him from the more powerful guns preferred by villainous henchmen: it would take eight layers of laminate to stop a bullet from an AK-47.[19] The problem is not layering the glass but doing so in a way that you can still see through it to drive the car. The refractive index of the plastic must be very carefully matched with the glass to stop distortions or reduce transparency.

---

[19] The bullets still won't skim off the surface of the glass as they are seen to do in numerous Bond films.

It is not Bond's gun that saves him time and time again, but the skill with which he uses it, his apparent indifference to the risks of others firing at him, and his wits. Aboard Goldfinger's plane, Pussy Galore declines to shoot Bond after he tells her the likely consequences. In fact, Bond exaggerated. The .45 ammunition in Galore's Smith & Wesson has a particularly large diameter (0.45 of an inch, hence its name) and was designed for military use. The idea was to create maximum damage to the person fired at, and keep the bullet inside the target so it could not go on to harm any of your fellow soldiers. Even though Galore aims her gun at Bond's softest bits, his belly, the bullet is unlikely to make it all the way through him. Fired directly at the fuselage, the bullet would probably go through the few centimetres of aluminium, but it is doubtful they would both be sucked out of the plane (see chapter 020). Unlike Bond, Galore was in no danger had she decided to fire her weapon.

At the climax of *The Man with the Golden Gun*, Bond and Scaramanga are back to back, weapons out and ready to see who really is the best shot in a duel.[20] After 20 paces, Bond turns and Scaramanga is gone. Bond defeats his opponent in his tailor-made funhouse/ training centre using his wits, by pretending to be a mannequin,[21] rather than with his superior gun or marksmanship.

---

[20] Insert your own joke about men comparing their weapons here.
[21] Insert your own joke about Moore's wooden acting here.

# 010

# *The Spy Who Loved Me* and the parachute jump

James Bond has many qualities but boring is not one of them. To be fair, everything around him seems to conspire to be overly complicated and extraordinarily perilous. In 007's world there is no such thing as a routine day, an easy assignment, or a simple commute. An average working day might include at least one gun-fight, fist-fight or car chase, and often all three. He lurches from one spectacularly hazardous situation to another even more dangerous stunt, sometimes with such speed there is barely enough time for a plot.

We love to see Bond's life threatened in spectacular and unusual ways. From the safety of a cinema seat, bloated with popcorn and soft drinks, in danger of nothing worse than missing some of the action for a necessary bathroom break or the parking meter running out before the end credits, we marvel at his death-defying exploits. Bond is not the first fictional hero to perform feats of derring-do, but the film franchise has taken his physical exploits to Olympic levels: faster, higher, stronger.

The stunts in the 007 film franchise are all the more impressive because our hero has no superpowers and little in the way of CGI to save him. The filmmakers pride themselves on both their innovation and doing the stunts for real wherever possible. It may not be the principal actors involved, but there is almost always a real person, and not a mannequin, crashing, falling, skidding or holding on for dear life in whatever unlikely situation the screenwriters have come up with. The sky is quite literally the limit and often the setting for these spectacular stunts.

Gravity has played such a big part in the Bond franchise it deserves its own screen credit. It has pulled dozens of henchmen to their deaths off buildings, cliff edges, out of planes and into heavy-duty machinery. Bond is subject to the same force as his opponents but he often uses gravity's pull to his advantage. While the bad guys plunge to their messy deaths, Bond can usually find a rope, handhold or giant advertising banner to hang on to.

## Falling with style

In the opening sequence of *The Spy Who Loved Me*, Bond, pursued by assassins, skis off a cliff and into one of the greatest stunts ever captured on film. He hangs in the air just long enough for us to wonder how on earth he is going to get out of this one, before a Union Jack parachute billows out from his back to the triumphant notes of the James Bond theme.

The first person to ski off a cliff and save himself from a catastrophic collision with the ground by deploying a parachute was Rick Sylvester. He called it 'an outdoor mountain adventure', or a 'ski-base jump' rather than a stunt, and he first performed it in 1972.[1] The Bond producers read about Sylvester's adventure in a magazine article and decided to ask him to recreate it for their

---

[1] Off the top of El Capitan in Yosemite National Park, California.

next film. Sylvester was willing to do it but emphasised it was not going to be easy. But Bond films would not be Bond films if they took the easy option.

Sylvester said that Mount Asgard on Baffin Island in Canada was the perfect place to perform the feat because it had 1,700m (5,800ft) of near-vertical rock and a decent slope at the top to get up some speed on skis. The site, however, was only accessible by helicopter. He would need the landing site carefully checked and clearly marked out. He would also have to wait for the perfect weather so he wouldn't be blown off course or into the rock face by a gust of wind. The 007 team were as undeterred by any difficulties as their fictional agent would have been. Sylvester, a small crew and a lot of camera gear were packed off to Baffin Island.

After 10 long days of waiting, the weather cleared and Sylvester had just enough time to make the jump before the weather turned again.[2] Of the five cameras that had been trained on him, only one managed to keep track of him all the way from a little yellow figure hurtling towards the cliff edge to the unfurling of his Union Jack parachute.

Sylvester admitted afterwards that he 'sort of messed up. I had some delay in getting stable; the position one wants to be in, horizontal, stomach to the Earth. This may have been due to the suit and heavy-ish ski boots, as opposed to the baggy jumpsuits most common among parachutists. As a result I fell further than planned, further

---

[2] The change in conditions was so quick everyone scrambled back into the helicopter, leaving all but the most valuable camera gear behind.

than anyone expected me to, before deploying the parachute.' The footage also captured the sight of Sylvester's ski clipping the canopy as it opened. If it had torn the material it would have been disastrous.[3]

The stunt set yet another almost impossibly high standard for the 007 films. Before *The Spy Who Loved Me*, pre-title sequences had featured a fight, or an explosion, or a bit of exposition to set up the film. From now on they were expected to feature an incredible stunt *as well as* fights, explosions and a bit of exposition. You could no longer have Bond simply parachuting off a cliff. In the following film, *Moonraker*, he had to be thrown out of a plane without a parachute, fight the baddies on the way down *and* set up the plot for the rest of the film.

The premise for the *Moonraker* stunt is that Bond is flying home from a mission, only to discover he picked the wrong plane. Everyone on board – the pilot, the hostess and the indestructible henchman, Jaws – is in the employ of Bond's enemies, and those enemies want Bond dead. And, of course, they choose the most complicated means of execution available to them. Everyone, except Bond, has a parachute. Rather than waste bullets shooting Bond, the gun is used to shoot the cockpit controls. The pilot then exits the plane, followed by the parachuteless Bond, and then Jaws.[4]

---

[3] Sylvester returned to the Bond franchise in *For Your Eyes Only*. He doubled for Moore to do a 30m (100ft) fall from a cliff to be stopped by his climbing ropes. Moore had only 1.2m (4ft) to fall for the close-ups back in the studio but said it was still very painful.

[4] How a 7ft 2in man hid on the tiny aeroplane is never explained. Nor is what happened to the hostess. She had a parachute so presumably made it to the ground safely.

Gravity accelerates everyone downwards at the same rate. The only thing to hold them back is a lot of air. The molecules that make up air are tiny, but crash into enough of them and your progress towards the ground will be slowed. Spreading yourself out horizontally, belly downwards, will maximise your personal impact on these gases, but will not slow you enough to avoid catastrophic injuries when you land. A parachute will increase the surface area enormously, creating much more air resistance, and slowing the descent further. Bond needs a parachute if he is to survive and the only one he can see is on the back of the pilot below him. To catch up with the pilot he streamlines his body, minimising the area pushing through the air. A brief tussle and Bond has a parachute, and the pilot is left to accelerate to his inevitable death.

Bond just manages to clip the borrowed chute in place before Jaws, who has used the same streamlining trick, catches up with him. Before Jaws can sink his metal teeth into Bond's calf, Bond pulls his rip cord and the parachute deploys. Bond slows but Jaws continues to accelerate away from him. Jaws also pulls his rip cord but with his immense strength it just snaps and his parachute is useless. His landing is cushioned by a circus tent and the indestructible henchman survives to menace Bond again later in the film.

Calculations carried out by physicists Metin Tolan and Joachim Stolze show that this entire sequence of events could theoretically be possible if everyone jumped from a height of 6,000m (19,600ft). Instead it was filmed from a height of 3,000m (9,840ft) in short sections over five weeks, 88 jumps and with parachutes for everyone.[5] These parachutes had to be redesigned so they could be concealed under clothing. It was still a considerable risk, particularly for the one person we never see, the cameraman. Not only did he have to worry about the usual dangers associated with jumping out of an aeroplane, but he had a camera attached to his helmet. More redesigning was necessary to make the camera as lightweight as possible, and to slow the rate his parachute unfurled, so he wouldn't snap his neck in the sudden deceleration when he pulled the rip cord.

The franchise's love of aerial stunts has been demonstrated over and over again but with many new twists on a recurring theme. A similar idea of too many people and not enough parachutes was used in 2008's *Quantum of Solace*, but this time Bond does the honourable thing and shares with Camille. The extra weight is an issue, especially when Bond shifts his position so Camille lands on top of him, but the bigger problem is the very short distance they have to fall. May Day, or at least the stunt actor that doubled for Grace Jones in *A View to a Kill*, showed it is possible to parachute from just 244m (800ft), off the Eiffel tower, and survive. However, a specialist parachute that will deploy rapidly to a very large surface area is required to slow them sufficiently. In *Quantum of Solace*, Bond and Camille give their shared parachute as much opportunity as possible

---

[5] It was done by the US parachute team. Jake Lombard, with his uncanny resemblance to Roger Moore, doubled for 007. The tallest member of the team, 6ft 5in Ron Luginbill, doubled for Richard Kiel's Jaws.

to slow them down by forcing their plane to climb as high as it can before exploding and falling into a deep sinkhole.[6]

The other popular choice for the franchise is vehicle-based stunts: not just planes, but trains, automobiles and every other conceivable, and sometimes inconceivable, mode of transport has been pressed into action. The winning combination must therefore be a vehicle doing a stunt in the air. And the 007 filmmakers have perfected it.

## Magnificent men in their flying machines

In *GoldenEye*, instead of falling *out* of a plane, Bond drives a motorbike off a cliff to fall after a plane and climb *into* it. The stunt uses the same principles of gravity and drag from *Moonraker*, but in a different arrangement. On the ground, the motorbike helps Bond get up to speed faster, but in the air its complicated shape will only slow the rate of his descent and so he discards it as soon as he is airborne to streamline himself and catch up with the plane falling ahead of him. Ignoring the fact that the plane's engines are running and would be accelerating the plane, and that the wings would give it some lift, Bond might just be able to manage to catch up in 1,200m (4,000ft). Filmed at Tellistock Mountain in the Swiss Alps, the stunt meant Bond still had 1,451m (4,760ft) left to get inside the plane, take control and pull the nose up before hitting the ground. But, of course, it wasn't filmed in one sequence.

---

[6] Calculations by Tolan and Stolze show Bond and Camille would be travelling at about 35kph (22mph) when they hit the ground, which is survivable but likely to cause a lot of damage.

Instead a stunt actor drove off the cliff on a motorbike, fell about 300m (980ft) before deploying a hidden parachute and made no attempt to get inside the plane.

The general lack of parachutes in Bond's world means he often has to find other ways to make it to the ground safely. Conveniently, villains are sometimes transporting other vehicles in their crashing aircraft that Bond can use to escape. Gustav Graves has a handy helicopter on board his giant Antonov aircraft (*Die Another Day*), and a Jeep has been stashed among the drugs being transported in a Hercules cargo plane (*The Living Daylights*), which Bond can clamber into and use to drop to the ground and get away.

Vehicles shown out of their natural habitat, like a Jeep or a motorbike falling out of the sky, has become a recurring theme in the 007 franchise. With Bond in control, cars and motorbikes aren't just driven around skilfully on roads, or even roofs. Boats needn't be restricted to the surface of the water. Why confine yourself to two dimensions when there is a third readily available, stunt actors willing to take calculated risks, and experts to do those calculations for you?

In *Live and Let Die*, Bond is trying to escape Dr Kananga's clutches in a borrowed boat. The problem is the henchmen have bigger, faster boats and Bond is running out of water. So he propels his boat into the air and over a road, a car, a sheriff and another henchman in the process of being arrested. The boat jumps 33m (110ft), a record that stood for three years, Bond finds some clear water, the sheriff has his car destroyed by the boat chasing Bond, failing the same jump, and the henchman gets away to continue the chase.

Of course, the following film had to go one better. Instead of a boat jumping over a road, this time a car had to fly over a river *and* spin in mid-air before landing safely. In 1974's *The Man with the Golden Gun*, Bond is chasing Scaramanga, but Scaramanga is in his car on one side of a river, and Bond is in a borrowed car on the other side. A collapsed bridge is used as a ramp to launch Bond's car into the air, twist it through a full 360° rotation, and land it on the remains of the bridge on the other side.

These kinds of aerial spins had been done in motor shows in the US for several years, but they had never been done on film and over water. For the first time in cinema history a computer was drafted in to crunch the numbers and calculate the optimum speed that would give the car enough momentum to clear the river (62kph/39mph). The speed of the car would also influence how quickly it rotated in the air, so it had to match the speed needed to clear the river; speed also determined the angle of the ramps required to give the car an initial twist (50°). The car was modified to ensure the driver, engine and everything else were centrally aligned to maintain the centre of gravity as it rotated in the air. All the other seats and extraneous bits and pieces, including the roof that was replaced with cardboard, were removed to keep the weight down. The jump was done in one take.[7]

---

[7] *The World Is Not Enough* combined both car and boat stunts in a 360° boat jump over another boat. Travelling at 70mph, if the boat had landed upside down it would have taken stunt driver W. J. Milligan's head off. Weeks were spent perfecting the stunt so that it didn't.

Bond – well, the stunt drivers in the 007 films – can make cars, boats and bikes do extraordinary things. Chasing after or away from bad guys might be considered exceptional circumstances, but there is rarely anything conventional about how Bond travels. Public transport is no exception. When most people choose to stay inside the transportation in their seats, Bond is usually clinging on for dear life on the outside.

## Hanging in there

The 1983 film *Octopussy* sees the villains and a bomb heading off in a circus train. Bond must follow, even if it means travelling al fresco. What looks like a relatively simple stunt, by 007 standards at any rate, shows just how dangerous this work can be. Martin Grace, doubling for Roger Moore, was on the side of Octopussy's train. The track had been checked for any obstructions, but for technical reasons the train overshot its mark and carried on into a section of unchecked line. Grace smashed into a piece of concrete jutting out towards the track, shattering his pelvis. Despite the horrific injury, he had to continue holding on until the train stopped or he would have fallen under the wheels to his death.[8]

It was the same stunt actor holding on to the side of a cable car hundreds of feet above the ground to double for Moore in Bond's fight with Jaws in *Moonraker*. Obviously there were safety cables so the stunt actor wouldn't fall but, due to a miscommunication, filming started before Grace could clip it on. 'I wasn't

---

[8] It took him six months to recover from his injuries.

attached by anything other than just pure terror.' He desperately held on to the handrail and waited for the word 'Cut!'

The aim in most of these precipitous situations is not to fall off, but occasionally that is the goal. It makes sense to either slow your progress towards the ground or cushion your fall when you get there. In *GoldenEye*, the villains have been inconsiderate enough to build their chemical weapons cache inside a dam. Bond decides to take the direct approach to gain access and jumps off the dam. With a bungee cord attached to his ankles, he launches himself off the 220m (720ft) Verzasca Dam in Switzerland.[9]

Bond's mission has clearly been carefully planned. The height of the dam is known, as is Bond's mass, so the correct bungee cord can be chosen. As well as choosing the right length of cord so Bond can comfortably reach his target, the material also has to be carefully tailored to the situation. After falling several hundred feet, Bond will have built up a lot of momentum. Stopping suddenly would cause huge forces to pull at his body, causing injury. Slowing him down has to be done gradually. Composed of many latex strands within a protective sheath, the bungee cord will stretch to absorb the energy Bond has built up. Too stretchy and not enough energy will be absorbed; too stiff and Bond will stop too suddenly; too weak and the cord will stretch beyond its elastic limit and snap.

---

[9] Though it was Wayne Michaels, not Pierce Brosnan, who jumped. It set a world record for the highest bungee jump from a fixed point.

Sometimes these stunts have also stretched the audience's credibility to its elastic limit but rarely past breaking point. We can easily forgive clever editing to slice different bits of footage together to make it look like one continuous sequence. No one minds a little CGI to erase safety ropes, when it is clearly real people going to considerable risk to do these stunts for real, or as near real as safety will allow. It is all the more impressive that the Bond films go that extra mile to create not just new and exciting stunts but to execute them with such style. It is a testament to the stunt coordinators, stunt actors and technicians who make Bond look as good as he does.

## Pulling it off

Every 007 film contains many, many stunts, far too many to go into in one book, let alone one chapter. Dozens of cars, planes, boats, bikes and helicopters have been wrecked in the process. The franchise has kept a lot of British stunt actors in work – over a hundred of them were needed in just one film, *You Only Live Twice*, when dozens of ninjas are seen abseiling from the roof of Blofeld's volcano lair.

The films are so stuffed full of action, some of which may seem so low-key or commonplace that it doesn't qualify as a stunt, but it still poses a real risk to the actors involved. Bond has thrown and absorbed so many punches over the decades it is easy to forget these fights are carefully choreographed, and that small mistakes and slight deviations can result in injury. George Lazenby and Daniel Craig have both broken colleagues' noses.[10]

---

[10] It happened during a screen test, and Lazenby's commitment to the fight was part of the reason he was cast.

Grace Jones was similarly enthusiastic when strangling David Yip in *A View to a Kill*. It took three takes for him to be murdered to the filmmakers' satisfaction. Yip didn't think he would have survived a fourth.

Killing off characters is to be expected but killing your actors is to be avoided at all costs. Obviously, everything is made as safe as possible when filming. Mock-up sets are built for stunt actors to rehearse fights. When things are to be broken, lighter, safer versions are used. Wherever possible, sugar glass is substituted for the real stuff because you can't really cut yourself on it.[11] However, the glass floor in *Die Another Day* had to be real to support the stunt actors' weight. When the actors fell through it they knew they would get cut. All stunt actors know that, sooner or later, they will get hurt. And for the bigger stunts some don't eat beforehand in case it goes wrong and they need to be operated on.

However meticulous the planning and careful the choreography, accidents are inevitable. Harpoons have gone through legs and emergency bottles of oxygen have slipped out of reach in underwater scenes (*Thunderball*), explosions have been mistimed, setting fire to actors (Roger Moore in *The Spy Who Loved Me* and Pierce Brosnan in *The World Is Not Enough*), and people have slipped and tripped (too many films to list). All the Bond actors have sustained injuries during filming.[12]

---

[11] Modern films may use tempered glass that is designed to shatter completely into rounded lumps rather than large, jagged pieces that are more likely to cut someone.

[12] Roger Moore would drolly complain that the filmmakers were trying to kill him as they would make more money out of the insurance than box office receipts.

The stunt crews have been cut, bruised, bashed and left with broken bones. Incredibly, more serious accidents have been very rare in the Bond films. Aerial cameraman Jonny Jordan lost a foot when filming the helicopter battle in *You Only Live Twice*. An updraft brought the rotors of one helicopter into the path of the helicopter Jordan was filming from.[13] Stunt work is dangerous. Stunt actors deserve all the admiration, respect and extra pay they receive for taking on these risks.

As more and more 007 films have been made, bigger stunts, more dangerous situations and new combinations of vehicles doing amazing things have been devised. Where will all this lead? Presumably as far as the filmmakers' imaginations, the laws of physics and the insurance premiums will allow.

---

[13] He returned to filming with a prosthetic foot.

# 011

# *Moonraker* and the exploding space station

If I can guarantee one thing in a Bond film, it's that something, probably a lot of things, will explode. From *Dr No* onwards, the 007 filmmakers have taken every opportunity to blow up anything and everything with an enthusiasm and flare/flair that is impressive to say the least. Anything that can explode certainly will, and sometimes things that have no obvious reason to do so still do.

You might think that after 25 films, seeing things burst into flames would get a little dull and repetitive, but fireballs and big bangs have an enduring fascination for most people. And the filmmakers have not just repeated their explosions but have found new and creative ways to show bigger and better balls of flame billowing across cinema screens. The films stray very far from the path in terms of realism, but the visual effects they produce are phenomenal.

*Moonraker* has some of the most ambitious sets and set pieces of the series. Bond's nemesis, the laid-back Hugo Drax, doesn't want mere money or power, he wants to set himself up as a de facto god. He has built himself a secret space station from where he can look down on Earth as he destroys all of its human inhabitants, ready to replace them with his band of beautiful people. Of course, Bond manages to sneak aboard and, with the invaluable help of CIA agent Dr Holly Goodhead, sabotages the whole operation. Bond and Goodhead only just manage to climb into a shuttle and escape before the space station explodes, as all secret lairs should, in an enormous ball of orange flames. It makes little sense scientifically, but it looks great.

## Having a blast

Simply put, an explosion occurs when a large amount of energy is released very quickly. It can happen in many different ways, as illustrated in the 007 franchise, but all of these different explosions can be grouped into three main types: physical explosions, from the build–up of pressure; chemical explosions, where energy is released from chemical reactions between atoms and molecules; and atomic explosions, where the energy comes from within the atoms (discussed in chapter 013).

Chemical reactions involve breaking and making bonds between atoms. Molecules in less stable arrangements can be nudged to reorganise their atoms into more stable configurations. Weaker bonds are broken and stronger bonds formed, releasing a lot of energy in the process. Certain arrangements are known to undergo these reorganisations very rapidly. For example, gunpowder, or black powder, the dominant explosive material for centuries, is a mixture of finely ground carbon, sulphur and potassium nitrate ($KNO_3$).[1] A trigger, such as a spark, can break off the oxygen atoms from the potassium nitrate, which can then react with the carbon and sulphur to produce carbon dioxide ($CO_2$), sulphur dioxide ($SO_2$) and a lot of energy.

The nitrogen atoms that hold on to the oxygen in the potassium nitrate molecules are not just placeholders, they

---

[1] Gunpowder was discovered by the Chinese in the third century BC but not introduced into Europe until the thirteenth century AD.

can also react with each other to form nitrogen gas $(N_2)$.[2] This gas, like the carbon dioxide that is also released, takes up a lot more volume than the original solid or liquid explosive, so there is a rapid expansion. The energy released from the chemical reaction heats up the gases so they expand. Packing an explosive into a container means these gases are trapped, and pressure builds up until the walls of the container are ripped apart, creating the shock wave that is the source of the loud bang we hear.

The individual components of black powder must be ground finely and mixed very well to get everything as close together as possible, giving the oxygen quick access to the carbon and sulphur. It's the same idea behind burning fuels to release heat but, if it happens quickly enough, you get an explosion. Bonding all the necessary elements – carbon, hydrogen, oxygen and nitrogen – together on the same molecule brings them even closer to each other, and they can therefore react even more rapidly, producing more powerful explosions.

An ideal explosive material contains just enough, or even more, oxygen than is required to completely react with all the other elements within the material, so no oxygen need be drafted in from the air to fuel the explosion. The break-up of large molecules into lots of smaller ones also makes a more disordered system, and this releases yet more energy.

All that is needed to trigger the whole chaotic process is enough energy to break off the first few oxygen atoms. This energy can come from heat, an electric current or the percussive force of something hitting the explosive

---

[2] This is a massive simplification of what is a chemically very messy process, but the point is atoms swap partners to release energy.

material. The energy released from a few oxygen atoms reacting breaks off more oxygen atoms to carry out more reactions, and so on and so on. The whole thing can be over and done with in a fraction of a second.

Carbon, hydrogen, nitrogen and oxygen can be arranged in a fantastic number of different ways to give enormous flexibility over the circumstances that will trigger an explosion, the amount of energy that will be released and how quickly. While weakness can be a virtue in an explosive molecule, chemical fragility can be taken too far. No one wants a slight knock or a warm day to trigger an explosion. Bond needs his explosives to fire when the timer triggers the detonator and not by a punch from a henchman or when he is jolted along a bumpy road chasing after a bad guy. Nitroglycerine, for example, is a much more potent explosive than black powder, but it is also much more difficult to handle.[3] Immanuel Nobel developed a way of producing nitroglycerine in bulk, but it was so prone to exploding it was too dangerous to transport any distance.[4] Alfred Nobel, Immanuel's son, found a solution to the problem by mixing the nitroglycerine with Kieselguhr, an absorbent clay. He patented the mixture under the name of guhr dynamite and made a fortune.[5]

---

[3] Ascanio Sobrero, who discovered it in 1846, discontinued his investigations when nitroglycerine's explosive potential became obvious.

[4] Nitroglycerine plants had to be built near the quarrying or tunnelling site where it was to be used.

[5] It is this fortune that funds the annual Nobel prizes set up to assuage Nobel's guilt at producing something that had been used so destructively in conflicts.

Many other explosive arrangements of the four key elements have been produced since nitroglycerine, and some, like trinitrotoluene or TNT,[6] have even become household names. Though explosives have many peacetime applications, war has always been a spur to seek substances and formulations that can create bigger bangs in a more controlled way. The Second World War saw the development of new explosive compounds and mixed them with gel-like liquids to produce mouldable, or plastic, solids. By the time the 007 films went into production there were even more explosives for the special effects team to choose from.

In the opening sequence of *Goldfinger*, a flexible explosive has been packed into a long tube that Bond has wrapped round his waist, making it easier to carry as he swims up to a heroin factory on the Mexican coast. Audiences watching the film when it was released in 1964 may not have known what it was Bond was squeezing out of the tube like toothpaste but would have quickly learned from the massive explosion that happens just afterwards.

Explosions themselves may be chemical chaos with molecular fragments flying all over the place, but the overall process can be remarkably well controlled through design of the explosive material. More control can be introduced by using different explosive materials in combination.

---

[6] TNT was invented in 1863 by Julius Wilbrand, and it was originally used as a yellow dye. It was decades before people realised its explosive potential because it was more difficult to detonate than nitroglycerine. TNT is now the standard by which all other explosives are measured, the explosive force they produce known as the TNT equivalence.

There are usually two components to an explosive device: a primary explosive that is very easy to trigger but does not release as much energy, and a secondary explosive. The secondary explosive can release a lot more energy but it is more difficult to detonate, making it safe to store, transport and use. Despite what you might have seen on screen, such as in *Spectre* when Bond shoots a bomb in a briefcase causing it to explode, some secondary explosives are so stable that even a bullet being fired into them will not cause a detonation – that is what the primary explosive is for. Bond was clearly unlucky; his bullet must have hit the primary explosive in the film because what happens as a result is definitely an explosion. We see orange flames and lots of black smoke billow out from the building before it starts to collapse. It's what anyone might expect to happen except, like most explosions on screen, it's primarily for show.

### Fire!

Explosives can give you a lot of bang for your buck, but it's all over very quickly and there is not often much to see. The explosive material is used up so rapidly there is very little in the way of flames. Sure, debris will fly everywhere, there may be lots of dust and a brief flash of light. This is what we see in the opening sequence of *No Time to Die* when Vesper's tomb explodes as Bond is paying his respects. But it's the exception rather than the rule.[7] Giant balls of orange flames and black smoke is what audiences want to see and these are the result of something burning.

---

[7] There is also the nice touch of the shock wave damaging Bond's hearing, at least temporarily.

To satisfy cinema-goers, explosions need to last a bit longer and to look a bit more impressive. So that they can be seen more easily, flammable material, such as oil or petrol, is placed in such a way that it is ignited by the explosion and burns with the orange flames and dark smoke that look so good on screen. There is an awful lot of style that goes with the substance of a cinematic explosion, and the team behind the Bond films have mastered it. These flaming embellishments have gotten bigger with time, but the films started off credibly enough.

In *Dr No*, the titular villain needs a lot of power for his rocket-diverting radio transmitter and has therefore taken advantage of the uranium rocks that are apparently abundant on his swamp-infested private island and built a nuclear reactor.[8] When nuclear facilities fail, they can fail in a big way. Dr No has at least had warning signs dotted around the place and markers put on gauges to identify dangerous operating conditions. But Bond deliberately sabotages the plant to push the reactor beyond those bright red warning lines. Things start to go wrong fairly quickly, and continue to deteriorate, but at a pace that allows Bond and his love interest, Honey Ryder, to escape before everything is destroyed in balls of orange flames. However, this final, catastrophic explosion is not from a nuclear explosion.

The uranium used in nuclear reactors is not the right kind to generate a bomb (see chapter 013).[9] But if a nuclear reactor generates too much heat, the coolant

---

[8] I don't know how the radio transmitter works but let's just go with it.

[9] Presumably Dr No also has a refining facility on his private island.

will expand and pressure within the system will build. A pressure explosion will occur if the coolant becomes hot enough, which can cause the release of clouds of radioactive gas.[10] Yet this doesn't explain the smoke-filled corridors and orange flames Bond and Ryder must run through to escape. Something must be burning. Early nuclear reactors used graphite to control the nuclear processes within the reactor core, and graphite can certainly burn, but not so easily as the flames we see on the screen would suggest. Dr No must have a lot of other flammables stockpiled around the place. Perhaps he overestimated how much fuel he would need for his flame-throwing dragon tank.

Hoarding flammables would become another Bond villain trait as predictable as a physical deformity or an unshakeable belief that their plans cannot fail. Contingency planning is not their forte, except when it comes to fuel. In the follow-up film, *From Russia with Love*, this obsessive overcompensating leads to more explosions. Bond and Tatiana escape the bad guys by hopping into one of their boats, conveniently laden with several large barrels of fuel, more than enough to take them from the Istrian peninsula across the Adriatic Sea to Venice. When they are intercepted close to their destination, bullets fired towards their boat pierce the drums and show how little they have used – fuel pours out. The sight of the petrol spilling everywhere gives Bond an idea. He launches the barrels into the path of the enemy's boats and uses a flare to ignite the fuel spreading on the water's surface. The flames

---

[10] This is probably what happened in the Chernobyl disaster in 1986.

heat up the petrol still left in the barrels, the pressure builds until the barrels explode and ignite the rest of the fuel they contained. The result is exactly what you would expect in this situation: a lot of flames and smoke billowing everywhere.

Sometimes the enthusiasm for blowing up boats got the better of the filmmakers. When Largo's pride and joy, the *Disco Volante*, was destroyed in *Thunderball*, it didn't just explode, it was obliterated. The man in charge of the special effects, John Stears,[11] explained, 'I put the windows out 60 miles away. I was using deviants of rocket fuel and when we blew it I remember looking and there was a hole in the water of some 50 feet and the debris came out of the sky some three minutes afterwards.'[12]

The explosion of the *Disco Volante* may have been excessive, but boats do at least carry fuel to explain why they might blow up in the first place. Vehicles require fuel, so bright orange flames appearing after a car or plane crashes is sort of understandable, though designers and engineers go to considerable lengths to prevent vehicles from exploding if they are involved in a collision. Even petrol tankers have safety features to prevent the fuel they are carrying from exploding if they get into a crash, but perhaps the engineers didn't foresee the circumstances around the crashes in *Licence to Kill*. In this film, the villain Franz Sanchez is smuggling cocaine on a

---

[11] The real–life Q who made Bond's gadgets for many films. *Thunderball* won an Oscar for Best Visual Effects.

[12] You would need an awful lot of explosives to create enough pressure to break glass 60 miles away. Stears may have been exaggerating, but pressure waves can do strange things as they expand outwards.

massive scale by dissolving the drug in petroleum and transporting it by tanker (see chapter 016). The safety measures on these tankers are certainly up to withstanding most bumps and scrapes, but they're not usually being fired at with automatic guns and stinger missiles.[13]

The huge volumes of fuel in *Licence to Kill* are understandable because they are part of the plot. Having your secret base on an oil platform certainly makes it vulnerable to fires (*Diamonds Are Forever*). But why does Travelyan need so much fuel for his giant satellite dish hideout (*GoldenEye*)? Is he running everything off diesel generators? I suppose your secret lair needs to be off the mains to remain secret, but who delivers all that fuel to these remote locations and why stash it in such impractical places as the antenna suspended high above the dish? And what is Scaramanga doing with so many flammables about the place when his whole island operation runs off solar power (*The Man with the Golden Gun*)?[14]

*Quantum of Solace* at least offers an explanation as to why everything is so flammable in the desert hotel where the bad guys meet to do their bad-guy deals. In this case the hotel uses fuel cells to generate electricity from hydrogen. Hydrogen is of course very flammable but it burns with a pale flame and no smoke, unlike what we see on screen. But this is nitpicking. Bond films simply

---

[13] The tankers used in the film had to be specially made so as not to create shrapnel that could fly out from the explosion and hurt people. Each cost $100,000 and they needed 10 of them.

[14] The destruction of Scaramanga's island hideout involved skips – big skips – full of explosives. As Roger Moore and Britt Ekland ran from the explosions, Ekland fell back and Moore had to go back and drag her out of harm's way.

wouldn't be Bond films if we didn't see everything blowing up in huge balls of orange flame and clouds of black smoke, and the filmmakers have certainly come up with some inventive ways to produce those explosions.

## Getting creative

In *The Spy Who Loved Me*, Roger Moore's Bond disassembles a nuclear missile to retrieve the conventional explosive trigger that acts as the primary explosive in such weapons (see chapter 013). In *Skyfall*, Daniel Craig combines a few sticks of dynamite and some gas cannisters to create a much bigger bang than would be possible with either separately. But perhaps the most inventive explosion is in *The World Is Not Enough*. Pierce Brosnan's Bond deduces that an explosion is about to occur from a bit of fizzing on his fingertips. It's an extraordinary bit of chemical deduction.

In the 1999 film, Sir Robert King has left a huge stack of cash at MI6 headquarters for safekeeping. Unknown to him, and everyone else in the building, the notes have been soaked in urea and the metal anti-fraud strip in one of the bills has been replaced with magnesium. A signal from a transmitter hidden in King's lapel pin ignites the magnesium, the heat of which triggers the explosion of the urea. All the right ingredients are there, but they don't quite connect in the right way scientifically.

Because Bond had handled the money, some of the urea got on to his hands. When he adds ice to his whisky he notices a chemical reaction, though in reality the most that is likely to happen is that the urea would simply dissolve. He also seems to notice an unusual smell – perhaps some ammonia has been released from a chemical reaction between the urea and the water. This is possible but would not be accompanied by fizzing.

It is true that magnesium burns with a very intense flame that can be difficult to extinguish, making it a good potential detonator for an explosive compound. The problem is that the urea wouldn't explode. Urea is a common ingredient in fertiliser because it is an excellent source of nitrogen that helps plants grow. Fertiliser bombs are certainly a very real potential explosive, but it is not the urea that causes the explosion. Urea-soaked cash ignited by magnesium would undoubtedly cause a very nasty fire, and either the flames or some of the toxic fumes that would have been released could have killed King, which is after all the entire aim of this elaborate plot. But an explosion that blasts a hole in the side of the building? No.

To get from a bit of fizzing and an odd smell to realising a stack of banknotes is about to explode is several big intellectual leaps. But then, James Bond wouldn't be the highly respected secret agent he is if he weren't able to spot something suspicious and put two and two together to make four so quickly. In any case it is a great opening to the film.

Most 007 films rely on tried and tested explosive materials, and it is the situation that is unusual and the size that is all-important. *Spectre* went one better than all previous Bond films by blowing up Blofeld's desert lair. It may have been the biggest explosion ever featured in a film,[15] but within the context of the story, the blowing up of Drax's secret space station in *Moonraker* is much bigger.

---

[15] Detonating the leftover explosives from the shot apparently caused an explosion even bigger than the one caught on camera.

## Size matters

It's an indication of the scale of a Bond villain's self-delusion that they put so much effort into constructing elaborate lairs in such impractical places when even a slight acquaintance with the world of 007 would indicate how unlikely it is to remain intact. Drax may have thought his space station, hundreds of kilometres above the Earth, would be out of reach. But it isn't just Bond and Goodhead who manage to sneak aboard. Dozens of space marines soon hove into view to ruin Drax's day and destroy his new home.

Space is a vacuum, but inside the station the pressure must be kept close to Earth's atmospheric pressure so the inhabitants can breathe. There is a big difference between these two pressures and so any breach in the outer fabric of the station allows an explosive release of pressure from inside the station to outside. To achieve this effect, the filmmakers hid behind black screens forming the backdrop of Derek Medding's incredible model of the space station, and then took potshots at it with air rifles. Slow-motion footage of the debris blasted away by the pellets' impacts gives a fantastic impression of the gas exploding out from within. This is how a space station might be expected to explode; it is only the final shot of the entire structure disappearing in orange flames that is a stretch.

Quite apart from why Drax would have that much flammable material on board his space station, there is the problem of the oxygen needed to make it burn. There is no oxygen in space to help the process but it is obviously vital to keep the station's inhabitants alive. On the International Space Station (ISS), oxygen tanks are only kept as emergency reserves and the bulk of the inhabitants' needs come from oxygen released from water by electrolysis. Maybe Drax wasn't planning on staying on his station that long and just took lots of oxygen tanks with him rather than going to the trouble of installing an electrolysis system. Perhaps he decided to use flammable gas to propel his station into its gravity-mimicking spin. It would explain the visual effects of the station exploding, but not the sound. There is no sound in space, but we are so used to hearing and seeing explosions together that a silent explosion would be jarring and distract from the climax of the action.

The on-screen explosions in 007 films may have little scientific rationale, but the point of a Bond film is spectacle and entertainment. The scientific credibility may be stretched within the context of the film, but the scientific, engineering and technical skill of those making them is undeniably first class.

# 012

# *For Your Eyes Only* and electrocution through headphones

The Bond films don't set out to shock anyone, except when it comes to henchmen. Death by electrocution is a surprisingly common way for the bad guy's minions to meet their maker in the 007 franchise. It may seem like a fairly straightforward way to die, but it's surprisingly complicated, as are the methods that have been employed to deliver those lethal electric shocks in the films.

Large-scale evil plots require industrial-scale power supplies. I'm surprised more villains haven't diverted some of the current from their nuclear reactor or solar arrays through the furniture to deal with uncooperative or underperforming minions. If SPECTRE technicians can wire up a volcano it seems straightforward enough to connect a chair to the mains. It's got to be easier than maintaining a piranha pool. But, terms like 'straightforward' and 'easy' do not apply to the world of James Bond. Discovering new and unusual ways to use something as ordinary and everyday as electricity to kill people is all part of the fun.

How electricity is used as a weapon in the 007 films is emblematic of how the films operate as a whole. In the opening sequence of *For Your Eyes Only*, a helicopter has been rewired to deliver a lethal shock – not to Bond, no, that would be too obvious, but to the pilot. His headphones spark and he slumps over the controls. It would actually be an effective way of killing someone, but it would take so much effort to set up that only a Bond villain would bother. Where the master criminals meticulously plan their convoluted schemes, Bond is able to act on impulse and adapt to the situation. In the middle of a fight, if there is nothing else to hand, Bond will use any exposed wire he can grab to shock his opponent and give him the edge. Let's take a closer look at the deliberate, improvised and accidental electrocutions that have featured in so many Bond films.

## Electricity

Electricity is simply the movement of charged particles. Negatively charged electrons flow through metal wires and silicon-based circuits to power and control electronic devices. Inside our bodies, our nerves transmit their instructions via electrical signals generated by the movement of positively charged potassium and sodium ions.[1] The mechanisms may be very different but the principle is the same. It means that our bodies are vulnerable to electrical energy that can disrupt or overwhelm the normal electrical processes going on within us. And these bioelectric vulnerabilities have been thoroughly exploited by the makers of the 007 films.

The degree of damage done to a material by electricity depends on how much electrical energy actually flows through it and for how long. An indication of the

---

[1] There are also negatively charged chloride ions and lots of other chemical species carrying a charge, but for simplicity's sake we will stick with the potassium and sodium.

amount of charged particles being moved is given by the current, measured in amps (A). Materials resist the movement of these charged particles through them to a greater or lesser extent. How much resistance is measured in ohms ($\Omega$). To overcome resistance, charged particles are given a 'push', measured as the voltage (V), to force their way through. The number of particles, the push they are given and the resistance they encounter are all interrelated.

The electrical infrastructure inside our body, the nervous system, is a complex interconnected network of cells beautifully orchestrated to control almost every aspect of our life, from thinking, to moving, to monitoring our most basic functions. It is also marvellously efficient. There is little resistance to the flow of charged particles, thanks to our water-filled bodies, meaning individual nerves operate on tiny amounts of electrical energy – less than a tenth of a volt – but it also means that it requires very little electrical energy to potentially cause problems.

A shock of 0.03A to the hand will result in painful muscle contractions and consciousness is likely to be lost at around 0.04A. If the voltage is sufficient to force an alternating current of 0.05–0.08A through the heart, it can be fatal in seconds.[2] Increasing the current to just 0.1A can cause ventricular fibrillation and arrest in just 0.2 seconds. Ten times that amount,

---

[2] Alternating current (AC) is more dangerous than direct current (DC) because it is more likely to cause cardiac arrhythmias. A current of 0.25A DC to the heart is often survived, though this is not an invitation to try it out for yourself.

2A, and ventricular standstill (abrupt halting of the heartbeat) can occur.[3]

Most deaths from electricity are from cardiac arrhythmias, usually ventricular fibrillation ending in arrest. The second, but far less common, cause of death is from respiratory arrest. Overstimulation of the brainstem, and in particular the medullary respiratory centre of the brain that controls breathing, can kill quickly. Respiratory arrest can also occur when the current passes through the thorax, causing the intercostal muscles and diaphragm to go into spasm or become paralysed.

## Shocking

Some animals have taken advantage of the bioelectric capabilities of a nervous system by enhancing their own and using it to attack weaknesses in their prey. The electric eel has specialised electric organs, stacks of electrocytes (modified nerve cells) that act like individual batteries wired up together by a mass of nerves to combine their individual electric potential into much larger shocks. These animals can produce up to 860 V to kill their prey, shocks that can paralyse respiratory muscles even if they don't kill directly.[4]

Fatal fish are a mainstay of a villain's lair (see chapter 008), but in *Licence to Kill* they are not just protective pets. Milton Krest is a marine researcher with a warehouse

---

[3] A direct current above 4A may cause an arrhythmic heart to revert to a normal sinus rhythm, such as in a medical defibrillator.

[4] Despite their name, these animals aren't in fact eels but a species of knifefish.

full of exotic sealife that is a cover for his drug smuggling operation. Timothy Dalton's Bond goes for a nose around and bumps into a henchman, who he yanks into a huge fish tank containing an electric eel. The red and yellow hazard symbols and labels on the tank have warned us what to expect. A few flashes of light and a bit of thrashing around and it's all over. Electric eels can generate a spark, but the flashing lights are a little much. The action moves on and the shocked and stunned baddie is left to drown.

The water surrounding the eels and their victims helps conduct the electricity. Pure water is not a very good conductor of electricity, but the river water that is an electric eel's natural habitat, and the water in its tank in Krest's warehouse, won't be very pure. Nor will the water pouring into the mine in *A View to a Kill* or the Venice canals in *Casino Royale*, the settings for other electrical deaths in the franchise. Even a few soap suds can make a big difference.

In one of the many iconic scenes from the 1964 classic *Goldfinger*, Bond enters a girl's apartment just as she is getting out of the bath, distracting him from the fact that they are not alone. As Bond kisses her he notices a man reflected in the girl's eye and whirls round to face him. A fight ensues that ends up with the mystery assailant being pushed into the bath. He splashes around trying to reach Bond's gun, hanging in its holster from a nearby hook. Bond knocks an electric heater into the tub. There's a flash of light, a red glow and lots of steam. Bond quips 'Shocking,' as he calmly retrieves his gun from the now dead assailant, puts on his jacket and heads out the door towards the title sequence.

To prevent everyday electrocutions, such as from devices falling into the tub, metal objects are usually earthed. A thick piece of wire running from the pipework in the tub to the ground offers an easier route for the electricity to flow and minimising the amount that can affect anyone having a soak at the time, but it can still give a very nasty shock. Electricity will always take the path of least resistance, and wet skin offers very little resistance. Dry skin, however, is a completely different matter.

The electrically sensitive nerves within our bodies are usually protected by an almost complete covering of skin. Dry skin's resistance to the flow of electricity averages between 500 and 10,000$\Omega$ depending on the thickness of the keratin layer in the epidermis. Palm skin, where the keratin layer is much thicker, may have a resistance in the order of 1 million $\Omega$. The more the resistance, the higher the voltage needed to force the flow of charged particles through the skin and into the more conducting tissues below.[5]

There is a surprising amount of water sloshing about in Bond films, but not always when you need it. Bond's world is also stress-filled and often physically demanding, meaning a lot of the time, a lot of people are producing a lot of sweat. Sweat is produced by the body primarily to cool it down by the water evaporating from the skin. And though sweat is mostly water, it also has trace compounds and minerals that can help conduct an electric charge. It means the skin's normally high

---

[5] This is why the skin can show electrical burns when internal structures appear undamaged.

resistance is likely to be considerably lowered when Bond or the bad guys try to direct an electric current into a person's body, as they so often do.

For example, at the climax of *Goldfinger*, Bond and the formidable henchman Oddjob have their final showdown inside the locked vaults at Fort Knox. There's no water to be seen but there is lots and lots of metal, which is an even better conductor of electricity. Bond is bashed about and knocked against metal bars, floors, grills and staircases. He also has to dodge Oddjob's metal-rimmed hat, which conveniently slices through a major electric cable on the wall behind him. Another throw of the hat sees it embedded in the metal bars that guard the vault's gold. As Oddjob goes to retrieve it, Bond grabs the still sparking cable and presses the exposed wires to the metal bars. A complete circuit is momentarily created from the high-voltage cable, through the metal bars, the metal rim of the bowler hat, into Oddjob's hand, through his body, through his heart, down his legs and into the metal floor. Sweaty hands, as well as sweaty socks and shoes from the fight, would ease the flow of electricity through otherwise highly resistant material, like skin, hat felt and leather. The high voltage in the cable would also have helped force the current through any remaining high-resistance sections of the circuit. Oddjob is killed in a flurry of sparks and smoke.[6] The thick insulation around the cable protected Bond,

---

[6] The actor playing Oddjob, Harold Sakata, was injured by the sparks but clung on regardless to get the shot.

though he is lucky the electricity didn't arc to his hands and through his own body.

A convenient set of circumstances and Bond's quick thinking helps him defeat Oddjob. Where Bond adapts to the situation, Bond villains contrive the situation to their advantage. Someone as powerful as Blofeld can design his secret facilities to his own exacting and unusual specifications without attracting awkward questions from workmen. The fear he can induce in his staff also means he can get them to sweat almost on demand.

## Making a connection

The 1965 film *Thunderball* begins with SPECTRE associates sitting down for a meeting in their luxuriously appointed Paris offices. Apparently, even international crime syndicates have to go through the quarterly financial reports. In any other organisation, it would be a dull couple of hours talking about profits, projections and financial forecasts, but SPECTRE is not any other organisation. Agents Number 9 and Number 11 have failed to declare sufficient profits from their drug smuggling operation and one of them is suspected of the worst crime: thieving from the thieves. The repercussions are severe. Number 9 is unceremoniously electrocuted through his committee room chair. His smoking corpse is lowered into the floor and the meeting carries on. It is an execution, just like those carried out by states using the electric chair, though without the usual judicial process and with Ken Adam's stylistic touches.

Execution by electric chair originated in the US and was first used in 1890 in Auburn, New York in the belief, based on evidence of accidental electrocutions in the home, that it would be quicker and more humane than other forms of execution in use at the time. It soon replaced hanging as the foremost method of execution in the country.

Over time, every possible aid was used to ensure the electric current was directed towards the parts of the body that would cause the quickest death, i.e. the brain and the heart. Metal electrodes the shape of a skullcap were placed on the back of the head and the forehead. More electrodes were attached to the legs to encourage the flow of electricity through the body. The skin where the metal touched was shaved then moistened with salt water or gel to increase the conductance. The first 'jolt' was of 6–12A and 2,000–3,000V. It lasted for a few seconds before being switched off, whereupon the body relaxed and could be checked for a heartbeat. If the prisoner was still alive, another jolt was applied. The effects on the prisoner have been graphically described by Harold Hillman in his 1993 paper assessing the pain likely to be experienced in different forms of execution: 'The prisoner's hands grip the chair and there is violent movement of the limbs which may result in dislocations or fractures. The tissues swell. Micturition and defaecation occur. Steam or smoke rises and there is a smell of burning.'

Ken Adam's sleek design does not allow for practical features such as metal electrodes to protrude from the chairs at the SPECTRE meeting. There is no obvious metal contact with the skin at all, but the accusations from Blofeld will certainly have made Number 9 sweat, and he clearly receives a massive jolt. His body arches up and collapses back into his chair immediately. There is a bright flash of light, sparks and smoke. It's surprising how little the filmmakers sanitised Number 9's death. Later films are not quite so graphic and show more cartoonish electrocutions.

The opening sequence of *For Your Eyes Only* features a bald-headed man with a white cat trying to kill Bond. Bond is standing alone by a church, laying flowers on his wife's grave. A simple sniper shot from a safe distance and criminals and evil-doers around the world could be rid of the biggest thorn in their collective sides. But the Blofeld-like character, who is not named Blofeld for legal reasons,[7] is miles away on the roof of an abandoned building in east London, sitting in a wheelchair with a complex set of controls in front of him. Bond receives a summons back to HQ and climbs into the helicopter that has been sent to collect him, unaware it has been sabotaged by not-Blofeld. As the helicopter moves over the London skyline, the pilot's headphones spark and he slumps over the controls, dead from electrocution. Not-Blofeld takes control of the helicopter with his remote-control set-up and tries to kill Bond with some daredevil flying. It's an incredibly expensive, convoluted and public method of execution. There are easier ways to kill people, even people as resilient and resourceful as James Bond. But, even with the cartoon-blue, electric flashes around the headphones, the pilot's death is a

---

[7] Cubby Broccoli and EON productions lost the rights to both Blofeld and SPECTRE, meaning both disappeared from the franchise for many years.

more credible electrocution than the one shown in *Thunderball*.

If Blofeld has gone to all the trouble of rewiring a helicopter for remote control, he could easily have tampered with the pilot's headphones, exposing a few wires and connecting them to a high-voltage supply. The headphones, though they have insulating material around the actual earpieces, could easily have been sabotaged or simply be sweaty from being enclosed around the ears. The route between the two earpieces would take the current through the brain, where it could disrupt nerve signals controlling breathing and cause unconsciousness, even if it didn't kill the unfortunate pilot immediately.

Up to this point in the film franchise, deaths from electricity had been brought about by diverting a current from a conventional source through a henchman. It was only a matter of time until someone decided to design an electrical weapon. That time came in the 20th 007 film, *Die Another Day*. This film took references from all the previous films, including electrical deaths, and turned everything up to eleven.

The film starts off with 007 using a conventional device as an improvised unconventional weapon – a defibrillator. Bond is in a medical facility being treated after the torture he suffered in a North Korean prison. M wants him to continue treatment, but Bond wants to get back to work, so he decides to escape. Bond slows his heart rate until it appears his heart is asystole, or flatline – the most serious form of cardiac arrest because it is usually irreversible. The medical staff respond with CPR and an injection of atropine, a drug that increases heart rate. But then they pull out the paddles of a defibrillator and charge them to 300 joules.[8] Before they can shock Bond, his pulse rate returns to normal, he wakes up and pushes the paddles towards two of the medical staff, shocking them instead. They fall to the floor but seem to be otherwise OK.

The shocks from a defibrillator were a bad decision for several reasons. Defibrillators are used to correct irregular, rapid heart rhythms, almost the opposite of asystole. Apart from being the wrong procedure, the paddles also would not affect the medical staff in the way shown. Normally, both paddles would be applied to one person's chest, on top of specially conducting pads or gel, so that the electricity can flow as easily as possible between them and through the nerves controlling the heart to correct abnormal rhythms. A paddle per person, with nothing linking them, means there is nowhere for the electricity to flow except through the length of their body to the ground, which in this set-up seems unlikely mostly because of the poor connections. The current would have to get through layers of clothing to reach the skin before entering the body, both of which are poor conductors.

Electricity is used as a weapon again later in the film. The bad guy, Gustave Graves, has an electrified robot suit. It's a ridiculous get-up but could actually deliver more credible shocks to its victims than the defibrillator.

A glove from the suit is used to apply 100,000V to Agent Jinx's neck when she is discovered snooping

---

[8] Joules measure energy in any form – electrical, chemical or other.

around the villain's lair.[9] The blue sparks shown on screen are unlikely but a useful visual guide to let the audience know what is happening. The voltage is much higher than that which would be delivered by a defibrillator, or even a taser, which would be the closest real-life equivalent of Graves' weapon.

Tasers are gun-shaped devices that deploy two metal barbs designed to hook into the clothes or skin of the target. Wires connect the barbs back to the device so that current flows in a circuit from the device down one wire to the barb, through the body to the second barb and back to the device. A taser can deliver up to 50,000V of arcing voltage so that a discharge will cross up to 5cm (almost 2in) of air, ensuring a discharge will reach the skin even if the barb gets caught in clothing. Tasers are typically programmed to fire for five seconds, not long enough for a full discharge.

Rather than deploying long wires towards a target, the gloves of Graves' suit could have barbs in the fingertips. Taser barbs need to be at least 10cm (almost 4in) apart to deliver an incapacitating jolt, but that could easily be achieved across a handspan. When Graves' henchman, Zao, first uses the glove to shock Jinx, she falls unconscious. This does not normally happen with a taser but the glove has twice the voltage and is being applied to the skin of the neck, under which are some very important nerves that lead to the heart, lungs and brain.

---

[9] This is the kind of voltage that runs through the cables supplying electricity to heavy industry. The substations they emerge from are usually covered in yellow warning signs graphically showing the dangers of coming into contact with these kinds of voltages.

When Jinx revives she is given further shocks to her torso through her clothes. This time she remains conscious but it is obviously painful, which makes sense given the poorer connection and access to underlying nerves.[10]

Later in the film, Graves, wearing the complete suit, applies its full electrical force to his father, through his uniform, and he collapses into Graves' arms. The shock lasts longer than that given to Jinx, but it doesn't kill him or even knock him unconscious. Graves takes advantage of his father's weakened state to pull out a gun and shoot him. If the suit is powerful enough to deliver a jolt through layers of clothing, how is Graves not electrocuted himself?

The suit or glove's wearer could be protected with a layer of wire mesh within its fabric to act as a Faraday cage. Named after the brilliant nineteenth-century scientist and electrical pioneer Michael Faraday, these metal mesh cages allow current to flow through them rather than transferring the current to what is inside. They are extremely effective but, if it were me, I would want my robo suit to cover me completely to stop the high voltage arcing to any exposed skin, especially not my head where there are breaks in the skin, eyes and mouth that are moist and could therefore easily conduct the electricity to my brain. Graves, as you might expect for a Bond villain, is a bit more cavalier in his attitude to personal safety. He is lucky not to get a taste of his own medicine.

---

[10] Metal zippers and underwire in a bra can conduct electricity very well and often cause burns to the skin and increase the chances of the shock doing more serious damage in electrocutions.

However ridiculous the scenario, the many electrocutions in the 007 franchise are, on the whole, scientifically credible as a means of dispatching a henchman or several. A few embellishments here and there, such as flashing lights and blue sparks, are understandable additions to inform the audience as to what is going on. Electricity offers a lot of creative potential, which the filmmakers have certainly exploited, sometimes to shocking excess.

# 013

# *Octopussy* and the atomic bomb

Bond and the bomb have a long and intimate history. He has survived exposure to radioactive mud, been handcuffed to an atomic weapon, and has saved the world from a nuclear war several times over. It's been a considerable learning curve. He has progressed from waving a Geiger counter at some rocks to deactivating an atomic device while dressed as a clown. After almost four decades of repeatedly getting up close and personal with radioactive material, it's surprising 007 doesn't glow in the dark.[1]

The Bond films first appeared on screen at the height of the Cold War when the threat of mutually assured nuclear destruction was very real. As nuclear technology and weapons proliferated, knowledge and hardware could easily leak into Bond's fictional world. The filmmakers, and their fictional villains, could take advantage of the situation by keeping the films topical without being overtly political. Tensions between East and West could be exploited for the villain's own, ideologically independent ends.

Atomic weapons and power feature most prominently in the 1960s Bond films, when fear and excitement about the technology was at its peak. The titular villain

---

[1] I'm exaggerating, of course; Bond would not glow in the dark regardless of how much radiation he had been exposed to, but his level of exposure would be a health concern.

in *Dr No* (1962) has his own private nuclear reactor to power the system he uses to divert America's rockets; Auric Goldfinger in *Goldfinger* (1964) buys an atomic device to irradiate the gold in Fort Knox, and in *Thunderball* (1965) nuclear weapons are stolen and aimed at Miami.

As real-world politics changed, so too did the films. *The Spy Who Loved Me* (1977) has British and Soviet secret services working together to stop Stromberg from starting a nuclear war. In *Octopussy* (1983) it is one rogue Soviet general who steals a nuclear device to start a war his comrades would rather avoid. In *The World Is Not Enough* (1999) the Cold War is over and the process of dismantling the accumulated nuclear arsenals provides an opportunity for the villain to steal one of the devices being decommissioned.

## Destroyer of worlds

The incredible potential of atomic power was first realised in 1945, just eight years before the publication of the first James Bond novel. On 16 July that year, a little over 6kg (13lb) of plutonium was used to create an explosion the equivalent of 13,600 tonnes of TNT. One of the senior scientists behind the construction of the bomb was Robert Oppenheimer. As he watched the mushroom cloud climb 12km (7.5 miles) into the sky, a phrase from Hindu scripture *Bhagavad Gita* entered his head: 'Now I am become death, the destroyer of worlds.' Nothing would ever be the same again.

The bomb exploited the inherent instability of plutonium-239 atoms. All atoms are made up of subatomic particles: protons, neutrons and electrons.[2] The positively charged protons are crammed together inside the unimaginably tiny nucleus. Positive charges repel, and it's only the presence of neutrons that stops the protons from flying apart. The more positive protons there are, the more difficult it is to hold them together. Unstable nuclei rearrange themselves, or decay, to more stable configurations by spitting out particles and energy from their nucleus. This is radioactivity. The particles that are forcibly ejected from these unstable nuclei can cause a lot of damage, as can the energy that is released along with them.

Plutonium-239, with 94 protons and 145 neutrons in its nucleus, is particularly prone to falling apart, releasing neutrons and energy in the process. It can do this spontaneously, but the process can also be induced by firing neutrons at its nucleus. Get lots of plutonium-239 atoms together and the neutrons released by the decay of one atom can provoke other atoms into decaying, setting off a chain reaction. In a bomb, two portions of plutonium-239 are pushed together, or a single block is compressed, using a conventional explosive, to bring the critical number of unstable atoms together. Nuclear bombs and nuclear reactors use the same principle of a chain reaction but one is controlled and one isn't.

In the 1950s and 60s, nuclear power promised endless, efficient electricity supplies, and atomic science was

---

[2] As a chemist, please believe me when I say that electrons are very, very important in all sorts of ways, but for simplicity's sake we'll ignore them here.

often promoted as an exciting new adventure popularised in comics and even by toys. But the devastation atomic bombs could cause was also well known. The bombing of Hiroshima and Nagasaki in 1945 had horrified the world. The world of 007 reflected the excitement of nuclear science but also the danger.

## An agent in the atomic age

Nuclear energy is only secondary to the plot of the first Bond film, *Dr No*, but it still plays an important role. Curious as to why a collection of rocks from the mysterious Crab Key has created so much interest, Bond uses a Geiger counter to determine they are radioactive.[3] While investigating Crab Key, Bond and Honey Ryder enter a swamp, unaware of the dangers of this particularly radioactive part of the island. When they are captured, they are told to strip, and are scrubbed down in decontamination showers to remove the radioactive mud from their clothes and skin.

The damage radiation can cause a body depends on the type and time of exposure. A single radioactive particle could break a vital structure within a person's cells that is the trigger for cancerous growth, but there is a much greater chance it won't. But the more radiation a person is exposed to, the more likely it is that some of that radiation will cause serious damage. The higher the dose and the longer the exposure, the more damage can be done.

---

[3] The device gives a click when particles emitted from a decaying atomic nucleus hit a detector. The more clicks, the more particles and the greater the radiation levels.

There are several ways scientists quantify radioactivity, radiation and the potential danger it poses. Some measure the activity of a sample, and others relate to the effects of radiation on the body. For example, grays (Gy) is a measure of how much energy is deposited in the body by the radiation it absorbs. A dose of less than 1.5Gy is unlikely to shorten a person's lifespan. A dose above 8Gy and death will probably occur within months due to damage to the immune system, leaving victims vulnerable to even trivial infections. Doses in the range of 10–12Gy cause death in days from damage to the intestines.

Rapidly removing possible sources of contamination, such as radioactive swamp mud, reduces the time it can affect the body. As Bond and Ryder move through the showers in Dr No's lair the radiation levels are constantly monitored. Their readings go from 88 and 95 down to 8 and 5, which are deemed safe, though there are no indications of what these numbers signify. Whatever Dr No's staff are measuring doesn't really matter; the point is, Bond and Ryder were exposed to high levels of radiation but decontamination brought these down to safe levels. Why Dr No is suddenly so concerned with Bond and Ryder's health is more difficult to explain considering he was trying to kill them moments earlier.

This wildly varying attitude to personal welfare continues in Dr No's nuclear power plant.[4] All the staff get protective suits, there are signs and red lines warning

---

[4] Designed by the legendary Ken Adams, the set is largely fiction, but elements were certainly taken from real life. Adams had met with experts from Harwell and toured round a nuclear power plant for ideas.

of dangerous working conditions, but the whole plant can be sabotaged with the turn of a single wheel (see chapter 011). There are also no safety barriers preventing Dr No from ending up in his own cooling pond, where he meets his death. The nuclear reaction running out of control in *Dr No* is an accident. In *Goldfinger*, it is the intention.

The third film in the 007 franchise took Fleming's original plot to steal the US's gold reserves and updated it. Critics had pointed out that Goldfinger wouldn't be able to physically remove all the gold from Fort Knox. As Bond put it in the film, 'Sixty men would take 12 days to load it onto 200 trucks.' So the scriptwriters set about finding an alternative. It took months of research and weeks of writing, but what they came up with lacks only in a few details, and the plan itself is ingenious.

Instead of stealing the gold, Goldfinger will render it worthless, thereby massively inflating the value of his own gold. To achieve this he has obtained a 'dirty bomb', a nuclear device to be detonated inside Fort Knox. According to Goldfinger's calculations, the radiation will keep the gold and Fort Knox off limits to humans for decades, 58 years to be exact.

Any atomic bomb is dirty – the dispersal of radioactive material over a large area has damaging long-term effects, a fact illustrated by the 1956 John Wayne film *The Conqueror*. It was filmed in Utah, downwind of the site where 11 atomic tests had been carried out the year before. Soil from the site was taken back to the studio in Hollywood so it would match with the location filming. Over the following 30 years, 91 of the 220 cast members, not counting the Navajo Indians who had been brought in as extras, died of cancer. Among the deaths was Pedro Armendáriz, the Mexican actor who played Ali Kerim Bey in the 1963 Bond film *From Russia with Love*.[5]

Goldfinger has only one small device, but the effects of the blast alone will be significant. The heat generated by a nuclear explosion is intense, easily enough to vaporise the gold within Fort Knox, which would then condense on the interior walls, floor and ceiling, assuming the whole facility isn't destroyed in the blast. The explosion would cause atomic debris, including neutrons, to fly out in all directions. Some of those neutrons would smash into some of the nuclei of the gold atoms. A gold atom with an extra neutron is very unstable. In the days and weeks after the explosion, some of the irradiated Fort Knox gold would decay into liquid mercury. A portion of the US reserves would be permanently destroyed, or at least transmuted into a less valuable metal. To add to the problems, Goldfinger's bomb is described as a dirty one.

The purpose of a dirty bomb is to add even more damaging material that will create long-lasting radiation. It was proposed in 1950 by the physicist Leo Szilard, not as a serious recommendation but as a theoretical possibility.[6] Szilard mentioned cobalt as a possible candidate because, like the gold in Goldfinger's plan, it is also susceptible to picking up an extra neutron. The

---

[5] Knowing how ill he was, Armendáriz's scenes were filmed early and doubles were used where possible so he could be released from the production. He died before the film premiered.

[6] Nevertheless, several devices were built. One, a kiloton device, was built and tested by the British in 1957, but it was considered a failure and never repeated.

resulting cobalt-60 is unstable, decaying over a period of years, spitting out dangerous amounts of particles and energy until it turns into stable nickel. Detonating such a device inside Fort Knox would make it, and much of the land around it, a no-go area for decades until the radiation had fallen to safe levels.

Goldfinger's bomb also apparently contains iodine, but why is a mystery. If his bomb is based on plutonium, radioactive iodine-129 will be formed regardless, as it is produced when the plutonium atoms break apart. Adding extra, everyday iodine wouldn't make much difference. The iodine could easily be dispensed with and there would still be plenty of awful consequences from Goldfinger's bomb. In fact, he could dispense with the atomic bomb altogether and still succeed with his plan. An ordinary explosive device could be used to disperse pre-prepared radioactive cobalt-60 all over the Fort Knox gold, rendering it and the contents of the vaults untouchable for decades. But audiences just hearing Goldfinger is going to use an atomic bomb will instantly recognise the disastrous consequences, even if some of the details pass them by. And Goldfinger does like to show off, so an atomic device it is.

The filmmakers had no idea what a nuclear bomb looked like, and they could be fairly certain most of the audience wouldn't either, so they improvised. Bond is clearly also unfamiliar with the inner workings of an atomic device and stares in confusion at the multicoloured wires, spinning discs, switches and flashing lights. As the counter ticks down to zero, a hand intervenes and, with 007 seconds remaining, flicks the off switch. Bond, Fort Knox and thousands of lives are saved.

Goldfinger's plan, had it worked, would certainly have had a dramatic effect on gold prices and the population of Fort Knox. But the potential nuclear impact in the following 007 instalment, *Thunderball*, was considerably greater. This time, a stolen nuclear device is set to be detonated off the coast of Miami unless the world's governments comply with SPECTRE's demands.

Fleming's novel of the same story contains considerable information about the device in question and how it works. M tells Bond, 'the nose is full of ordinary TNT with the plutonium in the tail. Between the two there's a hole into which you screw some sort of detonator, a kind of plug. When the bomb hits, the TNT ignites the detonator and the detonator sets off the plutonium.' Without the detonator the plutonium won't explode, as M explains: 'Apparently even a direct drop, like the one from the B-47 over North Carolina in 1958, would only explode the TNT trigger to the thing. Not the plutonium.' Fleming, to give credibility to his stories as he often did, was citing a real incident.

A flight transporting a nuclear warhead was passing over Mars Bluff in South Carolina when the navigator noticed a warning light, indicating the bomb had not been docked properly. He checked the bomb but as he pushed himself up, he accidentally grabbed hold of the emergency release pin. Fortunately the plutonium core was stored in another part of the plane. Instead it was the conventional explosives that detonated when the bomb hit the ground, but it still injured six people and produced a crater 21m (70ft) wide.

Much of Fleming's detail is left out of the film version but all the audience really needs to know is that two weapons have been stolen that are capable of destroying a city. However, one piece of information they do provide is very useful. The weapons have been stolen from a Vulcan bomber. These are real aircraft that carried blue steel missiles, equipped with Red Snow nuclear warheads. The Red Snow was a thermonuclear[7] 1.8 megaton device, over a hundred times more powerful than the bomb dropped on Hiroshima and more than a thousand times as powerful as Goldfinger's dirty bomb.

Thermonuclear devices are designed to undergo nuclear fusion (forcing atomic nuclei together so they fuse), rather than nuclear fission (breaking nuclei apart). Fusion releases far more energy than fission and the devices were designed in stages so that exploding the first stage triggered a bigger explosion in the second stage etc., increasing their efficiency.[8]

Detonating a device like the Red Snow would create a fireball a mile wide. The crater would be around 300m (1,000ft) wide and 60m (200ft) deep. Everything within half a mile of the bomb would be completely destroyed. Four miles away from ground zero only foundations would remain. Miami and most of its residents would be obliterated. The mushroom cloud would grow 24km (15 miles) high and spread over 48km (30 miles) across. Most of the cloud's radioactive dirt and dust would fall on the surrounding area but some would be carried

---

[7] Often known as hydrogen or H-bombs because some designs include deuterium, a form of hydrogen.

[8] Fission bombs, like those detonated over Japan, have very poor efficiency because the plutonium they are made from is blown apart by the initial explosion and stops the chain reaction from progressing further.

thousands of miles by winds before settling and contaminating the ground it fell on. Anyone within the radius of the cloud would be exposed to lethal levels of radiation, dying within hours. Further out, the radiation would still kill but death from damage to the immune system would take days or weeks. At 257km (160 miles) out, the radiation would cause potentially fatal problems to nerves and the digestive system.

SPECTRE's plan is averted at the last minute. Bond and a team of US Navy divers intercept the villain and his henchmen transporting the bomb to the detonation site. A huge underwater fight results in the bomb being recovered.

After his encounter with Goldfinger and another atomic near miss in *Thunderball*, Bond makes an effort to get to know more about these devices and how they work. It clearly pays off, as his next direct encounter with a nuclear weapon, in *The Spy Who Loved Me*, shows Bond has learned enough to pull one apart to get at the detonator.[9]

## How Bond learned to stop worrying and love the bomb

The 1977 Bond film ratcheted up the nuclear threat once again. Instead of one device to wipe out a major city, this time the bad guy, Stromberg, wants to trigger the detonation of *all* the bombs in a nuclear war, the result of which would be a nuclear winter (see chapter

---

[9] One would have thought the crew of the US submarine would be better equipped to do this – it is their missile, after all – but they just stand around watching.

005). To carry out his plan, Stromberg has captured three submarines, two of which will be crewed by his own minions to fire two missiles, one Soviet and one American, towards the opposing countries to trigger a war. A missile from a third American submarine, presumably part of Stromberg's back-up plan, supplies the explosives that allow Bond to blow a hole into Stromberg's heavily fortified control room. Once inside, he redirects the missiles to destroy the hijacked subs.

All these experiences have added to Bond's knowledge of atomic weapons. So when he is face to face with yet another one in *Octopussy* – and without any experts, Navy SEALs or submariners around to help him – he is able to disarm the bomb himself, just in the nick of time.

*The Spy Who Loved Me* had seen the British and Soviet secret services collaborating in their effort to avert nuclear armageddon. The Cold War continued its slow thaw so, when *Octopussy* was made in 1983, the majority of the Soviets are shown as unwilling to engage in war with the West. It is the splendidly over-the-top General Orlov who decides to go it alone. He borrows a nuclear warhead and, in cahoots with a jewel smuggler, Kamal Khan, plans to sneak the stolen weapon into a US airbase in East Germany.[10] He predicts that when the device explodes it will be attributed to an accident with a US weapon. As a result, Europe will unilaterally dismantle their nuclear programme and the USSR will be able to invade virtually unopposed.

---

[10] I imagine nuclear warheads are not easy to slip past security, but presumably Orlov's superior rank and terrifying personality mean he is able to get hold of one without too many questions being asked.

Khan's jewel smuggling racket uses Octopussy's circus as a front (see chapter 007). The jewels are normally hidden in the base of the cannon used for the human cannonball act in the show's finale. When no one is looking, they are swapped for the nuclear device – an object about the size of a fire extinguisher – that has clearly been removed from its missile housing. The bomb may have been trimmed of some of its bulkier components to fit into the available hidey-hole, perhaps rendering it less powerful. But Orlov doesn't need a huge explosion, just one that has clearly come from an atomic weapon, to put his plan in motion. Nevertheless, Bond predicts the blast will kill thousands, and the resulting war will kill thousands, if not millions, more. As the prop cannon, complete with bomb, is wheeled into the centre ring in front of a packed audience, Bond bursts into the circus tent disguised as a clown. He manages to pull the detonator from the device, stopping the countdown at zero.

In another uncanny Bond coincidence, the same year *Octopussy* was released the world really did come perilously close to nuclear war. On 26 September alarms sounded at the Soviet early-warning system, reporting the launch of a US missile from Montana and then, to everyone's horror, warnings of more intercontinental ballistic missiles appeared on their machines. Stanislav Petrov, who was on duty that day, decided it was a fault in the system and did not launch Soviet missiles in retaliation. He was right, and his swift decision-making prevented World War Three.

Two years later, the USA and the USSR agreed to begin a programme of nuclear disarmament. It was the beginning of the end of the Cold War. The shift in international politics, and consequently international espionage, was heavily referenced in the Pierce Brosnan Bond era. The plots that had driven *The Spy Who Loved Me* and *Octopussy* were no longer credible. However, nuclear disarmament presented a new opportunity for a Bond villain to get hold of material for an atomic bomb.

In *The World Is Not Enough*, the baddie, Renard, steals the plutonium core from a missile in the process of being decommissioned. The plutonium is shown on screen as a dark grey lump cast into two hemispheres. Kept separate, the two halves of the sphere don't contain enough plutonium to start a chain reaction. Renard blows up one hemisphere to create a diversion but keeps the other half to insert into the nuclear reactor on board a stolen submarine. The addition of the extra, weapons-grade plutonium to the reactor core means there will be enough unstable nuclei in close proximity to each other to send the chain reaction into overdrive. The resulting explosion, we are told, will wipe Istanbul off the map and Renard's accomplice, Elektra, will gain complete control over all the oil pipelines into Europe.

Renard's stolen plutonium must be reshaped so it can be inserted into the reactor core. In reality, this is not easy. Plutonium atoms readily rearrange themselves with changes in temperature, going through brittle and plastic stages that are a nightmare to cast into different shapes. But in Bond's world, submarines come equipped with handy gadgets that can squeeze a plutonium hemisphere into a glowing blue rod, like radioactive toothpaste emerging from a tube.

Many people expect radioactive material to glow, and some do. A blue glow can be seen in the water around nuclear reactors, rather than the rods themselves, because of Cherenkov radiation. Charged particles travelling through a medium faster than light waves create a kind of optical sonic boom.[11] The film gets the details all wrong, but the blue glow serves as a visual shorthand for the audience to know Renard's freshly cast plutonium rod is very radioactive. Fortunately, Bond reaches Renard just as he is about to push the rod into the reactor. He uses a pressure hose to forcibly eject the rod out of the reactor, saving Istanbul, and into Renard's chest, killing him. It is the last real reference to radioactivity and nuclear threats in the film franchise and completes the circle that started 37 years earlier with another nuclear reactor in Dr No's lair. From then on, Bond villains would have to find other ways to threaten the world.

---

[11] The intensity of the light can be used in medical applications as well as to determine radiation levels in nuclear power plant settings.

# 014

# *A View to a Kill* and May Day

To say that women are integral to the 007 world is understating the obvious. From Shirley Bassey belting out the title song at the beginning, to the woman wrapped in Bond's embrace at the closing credits, they are always front and centre. Aside from James Bond, they are usually the most prominent part of the advertising for any new film. Images of glamorous women adorn the book jacket covers and film posters, regularly overshadowing even the villains. Sometimes they're not even a whole person, just a shapely outline or an anonymous pair of legs. Slowly, oh so very slowly, the women that inhabit Bond's world have developed a third dimension: more depth, more lines and more relevance.

In a crowd of stunningly attractive women, burdened by sexually suggestive names but rarely weighed down with anything approaching substantial clothing, some have stood out for being more than a beautiful body. Not all of them are damsels in distress awaiting their knight in shining tuxedo to rescue them from the clutches of an evil villain. Sometimes it is the women who save Bond, or threaten him, or outsmart him. May Day, in *A View to a Kill*, in contrast to all the women before her, appears on the promotional poster standing back to back with Bond, equally self-assured and equally competent. The film's tagline asks, 'Has Bond finally met his match?' And the answer is yes. The Bond women, like their relationship with the franchise, are more complicated than the silhouettes of their naked bodies in the opening credits of almost every film might suggest.

## First impressions

Ian Fleming's relationships with women, in print and in real life, leave a lot to be desired. His fictional women are functional, they provide sex and ensure the smooth running of Bond's life without any of the 'annoying habits' like wanting to settle down or change him in some way. They type his reports, tidy his home and fret about him when he isn't there, but never cramp his style. They have a non-committal commitment to Bond's best interests.

Fleming's, and Bond's, reputation for treating women as disposable objects was certainly not helped by the 007 film adaptations. Bond expert Henry Chandler counted 14 women who slept with Bond through 12 books, and 58 women in the first 20 films. With a handful of exceptions, the actors are equally disposable, with very few appearing in more than one film.[1] These women were cast, certainly in the early days, primarily from modelling photos. There was little concern for their acting ability, and in most cases their voices were dubbed,

---

[1] Lois Maxwell, Samantha Bond and Naomi Harris have all repeated their roles as Moneypenny; Judi Dench as M; Eunice Grayson as Silvia Trench in the first two films; Léa Seydoux as Madeleine in *Spectre* and *No Time to Die*. Martine Beswick appeared as one of the gypsy fighting girls in *From Russia with Love* and as Paula in *Thunderball*. Only Maud Adams has appeared in three Bond films in different roles (Octopussy in *Octopussy*, Andrea Anders in *The Man with the Golden Gun* and an uncredited extra in *A View to a Kill*).

reducing them quite literally to objects of desire. The treatment of most of the women in the Connery era of James Bond can be summarised by one moment in *Goldfinger*. Bond dismisses his masseuse, Dink, with a slap on the bottom and the words 'Man talk.'[2] Of course, not all the women, even in the early films, are so inconsequential.

The first film, *Dr No*, has Ursula Andress in the role of Honey Ryder emerging from the sea wearing a white bikini and knife belt, a scene that has become an iconic, much repeated moment in film history.[3] A little later in the film, in a scene not re-shown nearly so often, Ryder describes how she took revenge on the man who had raped her; she put a black widow spider under his mosquito net. 'A female, and they're the worst. It took him a whole week to die.'

What she says is true; the female is more deadly than the male because the fangs of male black widow spiders can't penetrate human skin. The venom from the females is a mix of nerve agents that can cause severe pain and lead to respiratory and cardiovascular collapse. However, the spiders only bite to defend their web and egg sac, and not all bites are fatal, though they can take days or weeks to recover from. Ryder may not have intended to kill the man.

---

[2] Dink was played by Margaret Nolan, who also got the dubious honour of being a projector screen for the film's opening credits – a staggeringly on-the-nose metaphor for Bond girls if ever there was one.

[3] In the novel she is wearing only the belt and the knife, and Fleming describes Ryder as having a broken nose and a boy's bottom. Make of that what you will.

The message is clear: Ryder is a resourceful woman who is not to be messed with. It's a shame that this message is undermined by subsequent events. Up to this point she has been visiting Dr No's island without incident, evading his guards and collecting shells quite happily. After bumping into Bond, Ryder is shot at, chased by dogs, exposed to dangerous levels of radiation, drugged and almost drowned. Though Ryder starts off as the more knowledgeable member of the party, Bond soon dominates. Like some strange superpower, Bond's presence seems to drain women of rational thought and independent action.[4] Finally, Bond rescues Ryder from a situation that he was largely responsible for creating in the first place.

*Dr No* also introduces another iconic woman to the 007 series, though a woman who is strikingly more in control of her own destiny and the destiny of others. Never mind Blofeld or Goldfinger, Miss Moneypenny is the one who wields real power in Bond's world. George Lazenby's Bond describes her as 'Britain's last line of defence'.

Moneypenny is the real authority in MI6. She controls M's diary and decides who has access to his office. She is frequently in possession of more information than either Bond or M and covers for Bond when he deviates from M's strict orders. Moneypenny is also confident enough to rebuke Bond if he oversteps the mark. Their relationship is flirtatious but respectful.

---

[4] To be fair, this superpower also works on men, particularly arch villains who can run complex criminal networks with ruthless efficiency but always forget to shoot Bond in the head as soon as they come face to face with the spy.

If there is a real-life equivalent of Moneypenny it is, perhaps, Fleming himself. It was Fleming who sat outside the office of the chief of Naval Intelligence, filtering information, writing reports and guarding his boss's time from unwelcome interruptions. On screen, Moneypenny is portrayed as a little in love with Bond and slightly jealous of his romantic and international adventures. Similarly, Fleming idolised Bond, but as the man he would like to have been and living the life he would like to have lived. Moneypenny is the exception: a constant, assured, no-nonsense presence in an ever-rotating merry-go-round of women.

The filmmakers have gone to great pains to promote the Bond 'girls' as modern, independent and liberated women. But it's an independence that appears to extend mostly to their willingness to go to bed with a man without wanting any kind of commitment from him. It is certainly a convenient form of independence as far as Bond and the filmmakers are concerned.[5] These women may have important information to impart, but their main function is to occupy Bond's bed rather than advance the plot. Sometimes that amounts to the same thing. Tatiana Romanova is central to SPECTRE's scheme in *From Russia with Love* and her mission is to seduce Bond.[6]

---

[5] It may also be convenient for the women. Perhaps the reason so few of them stick around is because Bond doesn't live up to expectations.

[6] Fleming modelled Romanova on the real-life Anna Kutusova. On 25 January 1933, Kutusova, secretary at the British engineering firm Metropolitan-Vickers, was grabbed and bundled into a waiting car. It was the start of the investigation that led to the arrest of six employees on charges of espionage and spying (see chapter 002).

The other prominent woman in the film is Rosa Klebb, the villain of the piece. She is deliberately portrayed as unattractive, and a threat to Bond in more ways than one. She plans Bond's death and also tries to seduce Romanova.[7] Romanova only gets to act on her own initiative at the very end of the film, when Klebb attacks Bond and tries to stab him with her poison-tipped shoe. Romanova, a KGB agent, can shoot Bond, the enemy spy, or Klebb, her boss, compatriot and sexual harasser. She shoots Klebb.[8]

The following film, *Goldfinger*, marked a major shift in casting. Honor Blackman as the lead woman was older than her co-star Sean Connery and already had a considerable acting CV. In her role as Cathy Gale in the UK TV show *The Avengers*, she had proved her ability to thwart sinister plots and fight off bad guys. Blackman was an obvious choice if you wanted to show a similarly competent and strong woman in a Bond film, but Pussy Galore is not necessarily the role to do it with. There's the name, for a start. At least Galore is in charge of her own flying school, and able to fend off Goldfinger's advances with no more effort than a withering look. Sadly, Blackman's well-known judo skills were only put to use to fall safely when Bond throws her into a pile of straw. Her conversion to working for the good guys, by Bond forcibly having sex with her, is the worst example of a situation that occurs depressingly frequently in the series. Galore ends up doing all the work and Bond is

---

[7] Unlike in the novel, the film shows a rather half-hearted attempt.
[8] However beautiful, loyal and good Romanova may be, Klebb is the character everyone loves and remembers.

awarded the medal at the end of the film. And even her excellent flying skills are somewhat undermined when Bond must save them both from crashing in the plane he carelessly shot a hole in moments earlier. Having said all of that, Pussy Galore still marks a notable high point in the sparsely populated category of bold Bond women, and Blackman is fantastic in the role.

## Learning from past mistakes?

The follow-up to *Goldfinger*, *Thunderball*, acknowledged that Bond had not behaved well towards Galore, but then he goes on to behave even worse towards another woman. In this film, Fiona Volpe is bold and beautiful and works for the bad guys, just like Galore. But she refuses to convert to the good side, even after sleeping with Bond. Volpe spells out Bond's sexist assumptions about his influence over women. And just to give the knife a little twist, calls his sexual efforts with her a failure. Of course it isn't a failure. Bond can't fail. Volpe is killed just a few scenes later. She dies when Bond uses her body as a shield to protect himself from the bullets being fired by her accomplices.

Terence Young, director of *Thunderball*,[9] claimed the women in Bond were 'freer and able to make love when they want to without worrying about it.' In fact, they have a lot to worry about, choosing the wrong person to sleep with can, and often does, prove fatal. By the fifth 007 film, it was being spelled out to screenwriters as part of the winning Bond formula. Roald Dahl, who wrote the screenplay for *You Only Live Twice*, was allegedly told

---

[9] He also directed *Dr No* and *From Russia with Love*.

to include exactly three girls. The first should be on
Bond's side but 'bumped off by the enemy' about a third
of the way into the film. The second, working for the
enemy, was there for the middle third of the film and
should be brought over to Bond's side by his 'sheer
sexual magnetism'. If not, she should also be killed,
'preferably in an original fashion.' The third girl was
then to be brought in to support Bond through to the
closing credits where, having managed to resist his
charms for the last third of the film, she finally capitulates.

Dahl dutifully wrote parts for three Bond women, all
fitting the formula but none relevant to advancing the
plot, and all completely forgettable. The more memorable
Bond women are those who don't fit the formula, as *On
Her Majesty's Secret Service* would prove.

If one lesson had been learned over five films, it was
that actresses from *The Avengers* are a fantastic addition to
any Bond film. Following in Honor Blackman's heels is
Diana Rigg as Tracy di Vincenzo, the first woman to
tempt Bond away from his true love, M, and down the
aisle. It might not seem difficult to convince someone to
quit a high-stress, high-risk job, but only one woman has
managed it with Bond.[10] The marriage offer comes
with a large financial reward and the opportunity to
avenge a long-standing feud with Blofeld as a considerable
sweetener to the deal, but Bond makes it clear he is
marrying for love.

---

[10] Vesper Lynd in *Casino Royale* comes close, with Bond emailing
his resignation so he can live with her, just before she is forced to
betray him and takes her own life.

Di Vincenzo is a fantastic match for Bond. She is fearless, smart and resourceful. She out-skis, out-drives[11] and outwits all of Blofeld's hired heavies, impaling some of them on modern art sculptures along the way. She, like the Mounties, gets her man, and on her own terms. Di Vincenzo is brilliant, making her death at the end of the film all the more tragic. The 007 franchise sadly wouldn't see the likes of her again for a long time.

*On Her Majesty's Secret Service* was deemed something of a disappointment at the time of its release, and perhaps encouraged the franchise to fall back on familiar female tropes.[12] Beautiful but largely ineffectual women wearing very little continued to appear in the Bond movies and posters well into the 1970s.

## Charm offensive

Roger Moore's Bond charmed and tricked women into his bed rather than throwing them into piles of straw, but they were still shown as having little autonomy or competence. However, a concerted effort was made to portray a woman as an equal to Bond in 1977's *The Spy Who Loved Me*. When the call is made for the top Soviet spy to report for duty, it is not the hairy-chested man who is summoned but the gorgeous woman in bed beside him. Anya Amasova, Agent Triple X, is the KGB's

---

[11] Diana Rigg did some of her own driving, turning the cameraman green.

[12] I'm not suggesting Rigg's character was responsible for the drop in box office takings, or why Lazenby never reprised his role. There are many reasons why this film is and was considered a 007 outlier, and Lazenby could screw up his chances of a second film without Rigg's help.

equivalent to Bond. They work together for much of
the film, matching each other's spy skills along the way. If
one gets the upper hand, it is not long before the other
comes back with a better quip, idea or bit of information.
They are evenly matched right up until the end, when
Amasova is captured by the villain and tied to a chair – a
situation she is apparently incapable of getting herself
out of, despite the high levels of competence and
initiative she has shown over the preceding two hours of
film. Rescued by Bond from the sinking Atlantis, she
decides not to fulfil her promise of killing Bond in
revenge for him murdering her lover, and decides to
sleep with him instead. It's quite the turnaround
character-wise, but totally in keeping with how most of
the Bond films conclude.

Similarly, Dr Holly Goodhead in *Moonraker* is a CIA
agent and trained astronaut. She has all the gadgets, the
knowledge and the skills associated with Bond. She flies
the space shuttle while Bond shoots down the globes
filled with deadly poison. And of course, despite their
initial animosity, they end up in bed together. Bond
audiences grew accustomed to seeing women working
on a par with 007, but it couldn't prepare them for a
woman working against him and with every indication
of being able to defeat him. How could you prepare an
audience for the terrifying magnificence of Grace Jones
as May Day?

*A View to a Kill* is an updated *Goldfinger*. Instead of
Goldfinger trying to increase the value of his gold, Zorin
is trying to increase the value of his microchip stock.
Both villains are prepared to slaughter thousands to get
what they want and both are working in partnership

with a woman to achieve their goals. Goldfinger had Pussy Galore, and Zorin has May Day. And, though May Day sleeps with Bond, the roles are reversed. She is the dominant one and she is not persuaded to work for the good guys. May Day should die for her failure to observe the unwritten rules of Bond's world, and she does, but in a way that highlights other uncomfortable truths about Moore's 007. The promised fight between the two could never go ahead because the visibly ageing Moore was certainly no match for Jones. Instead May Day has an epiphany: she realises Zorin has no more regard for her than any of the other minions he has left to die in an explosives-filled mine. She moves the bomb away from the rest of the explosives and stays with it until it ticks down to detonation, killing her. Not only does May Day save Bond, but she also saves the day, and 007 is left literally dangling at the end of a rope. It was time for a new Bond.

May Day marks a notable change in Bond women. Up to this point, women had worked with and for the bad guys but only Klebb had been shown to be their equal. May Day is deadly *and* desirable, a trend the Brosnan-era Bond ran with.

**Women on top?**
May Day kills because people threaten her plans. Onatopp in *GoldenEye* kills because she enjoys it (see chapter 020). But Elektra King, in *The World Is Not Enough*, is willing to kill millions in a nuclear blast just to increase the value of her gas pipeline. She is the only woman to compete with other male Bond villains, such as Blofeld, Goldfinger and Drax, in terms of scale and

ambition. The size and severity of her potential crime
has to be enormous to justify her execution by Bond;
she is the only woman he has been directly responsible
for killing in 60 years.

The women in the films have been shot, stabbed,
drowned and suffocated by being covered in gold paint,
but always by the bad guys. In the weird morals of the
Bond films, Bond is the good guy because he doesn't
torture women, as Sanchez does in *Licence to Kill* when
he whips his girlfriend with the barbed tail of a stingray.[13]
Bond murders people but his killings are seen to be
justified because they prevent or atone for a greater
wrong. Not killing or beating women because they don't
do what you want is an astonishingly low bar to
overcome to be considered a good guy.

Brosnan's Bond not only had formidable female
villains to contend with but a female boss to answer to.
In this regard, as in much of its relationship with
women, the 007 franchise was behind the times.
*GoldenEye* hit cinemas in 1995, three years after Stella
Rimmington had been appointed as real-life Director
General of MI5.[14] And not even Dame Judi Dench is
immune to Bond's constant need to show his
superiority. Daniel Craig's Bond repeatedly hacks into
her computer and breaks into her home – with only a
verbal rebuke.

---

[13] A detail taken from Fleming's short story 'The Hildebrand
Rarity'.
[14] I know MI6 and MI5 aren't the same thing, but still. Another
high-profile appointment in 1992 was Barbara Mills as Director of
Public Prosecutions. Both appointments marked the first time a
woman had held those positions.

Pierce Brosnan's Bond is also paired with a fellow agent, as Moore is with Amasova and Goodhead. In *Tomorrow Never Dies*, the agent is Wai Lin of the Chinese secret service. Michelle Yeoh was already an established actress, known particularly for her roles in Hong Kong action films. She was the first Bond woman to get her own fight sequence, seeing off multiple attackers, with Bond arriving at the last minute to shoot the last attacker standing.[15]

Yeoh has a greater opportunity to show women are every bit Bond's equal than her predecessors. Bond's swan dive off a dam is daring; Wai Lin walking down a vertical wall in heels is class. She wields guns and gadgets with as much expertise and aplomb as Bond. She matches him stunt for stunt and at one point leaves him handcuffed to a drainpipe. But she is still the one who must be rescued by him at the end of the film.[16]

The women in the Bond films have certainly become more complex characters over time, with their own agency and with greater roles to play, but their relationship with Bond has not changed as much as we might like to think. Appearances are still everything. Their emergence from almost any situation with flawless make-up, hair and clothes is as improbable as the plots. It is surprising that fine female actors still accept these underwritten roles.

---

[15] Yeoh recruited some of her Hong Kong colleagues to film her fight sequence, something the UK stunt team was happy to hand over as the actors fought for real, albeit delivering softer punches and kicks.

[16] Resulting in the clichéd watery kiss that ends so many Bond films.

*No Time to Die* made a concerted effort to create better roles for women. In this film they operate on equal terms with Bond. Paloma, a CIA agent, works in partnership with Bond to extricate a kidnapped scientist and stolen weapon. Everything Bond does, she does but in a dress and heels. Another woman, Nomi, literally takes on his job as 007 and does it just as well, minus the fuss and personality clashes that characterise Bond's time in the role. Though it might seem a risk that these very capable women will show up Bond's inadequacies, they never do. Bond has been privileged to share the screen with some remarkable women and still be the hero.

# 015

# *The Living Daylights* and the cello case

In every 007 film James Bond confidently strides across the screen, turns suddenly and shoots a would-be assassin. The perfect timing, the measured, assured behaviour, is what we have come to expect from our hero. But sometimes he needs to pick up the pace a little bit.

Many of us work to deadlines, but in Bond's case the word has a very literal meaning. In his high-stakes world, time is of the essence and, understandably, 007 doesn't hang around. Contrary to common sense, he will head towards danger at full tilt, chasing down villains by whatever means necessary. But even our fearless secret agent has some sense of self-preservation. Occasionally he has to run away from the bad guys who are trying to kill him in some particularly unpleasant way.

You might think the idea of either chasing after villains or being chased by them would wear thin very quickly, but after 25 films the inevitable chase sequence is still something to look forward to. There is a lot to appreciate: the fast pace; the skill of those involved; the variety of vehicles that might be used; and whether Bond will be able to escape or catch up with his opponents, because his success is not a foregone conclusion.

As a man of so many well-honed skills, and with Q able to give any vehicle the edge over others, Bond could never be classed as an underdog. To allow his opponents a fighting chance, and to keep things interesting for the audience, Bond is regularly disadvantaged when it comes to chase sequences. His company car is not always available because he is overseas and relying on local transport; sometimes the self-destruct mechanism gets used before the chase begins (*For Your Eyes Only*), or the car is cut in half (*The World Is Not Enough*), or it isn't quite finished yet (*Spectre*). Q-car or no Q-car, he is usually outnumbered or out-horsepowered, and sometimes there is a giant space mirror melting the ice from under him (*Die Another Day*).

Bond is most successful when he is forced to improvise. In his world, cars, bikes, boats and buses are conveniently left parked up with a full tank and the keys in the ignition, ready to be commandeered by any international spy or henchman who happens to be passing. Almost every mode of transportation has been pressed into action, including those without an engine attached. Horses, skis, even a cello case will do in an emergency.

Even when 007 does have a gadget-laden car at his disposal, it doesn't always help. In fact, Bond is more likely to fail when he is driving one, and the heavily adapted and souped-up Aston Martin will be wrecked no matter what. It makes you wonder why Q-branch go to all that trouble.

### All mod cons

The Aston Martin DB5 from *Goldfinger* is the Bond car everyone loves. It is beautiful to look at and full of ingenious gadgets that genuinely help Bond evade the baddies on his tail. From the moment Tilly Masterson accidentally trips an alarm outside Goldfinger's lair, she and Bond are in a race to escape capture. The only worry is whether Bond will run out of gadgets before Goldfinger runs out of cars full of henchmen. The first car of gun-wielding guards is easily thrown off course with a cloud of smoke. Smoke bombs are comprised of material that doesn't burn very well, or where the rate of burning is controlled in some way, so particles of partially burned material are released into the air as smoke. Controlling the amount of oxygen that can reach the burning material, for instance by packing it into the brake light housing of a car, means it will smoulder rather than burn and so produce more smoke. The movement of the car will help the smoke billow out behind and blind any pursuing vehicles.

Soon after the first car is wrapped around a tree, another car full of henchmen appears in Bond's rearview mirror. Another gadget is deployed, squirting oil all over the road in the Aston's wake.[1] The oil and squirter shouldn't take up too much space or weight inside the DB5, as a little oil can spread very thin over a large area but still be effective at reducing friction and making the henchmen's car skid off a cliff and crash in a ball of flames.

When a third car hoves into view, Bond parks the DB5 and flicks a switch to raise a bulletproof shield that

---

[1] The original plan was to use tacks, but the filmmakers worried this would be too easy to replicate and switched to oil.

protects the rear window while he returns fire at the advancing baddies. You need a thick slab of tough metal to stop a bullet. Steel that has been heated and rolled to smooth out imperfections in the grain is a good choice. It's the sort of armour plating they cover tanks with. But it is very heavy. Just the weight of the metal alone would surely tip the Aston backwards on its bumper. And how big would the motor need to be to lift it? A metal like aluminium would be much lighter, though it wouldn't provide as much protection. Maybe it would have been enough for Bond to gain the upper hand if Oddjob hadn't shown up with his bowler hat. Tilly is killed as she tries to make a run for it. Bond, stooping to check on Tilly, is captured. However, the game is not over as Q-branch still has one more trick engineered into the Aston Martin.

An ejector seat is certainly a useful addition to a car, especially when your passenger has a gun pressed in your ribs. The special effects team built one into the car, though the compressed air system only had to fire a dummy out of the roof rather than a real person. But it adds yet more weight to the already overloaded vehicle, and should seriously hinder its performance. The more mass, the more force is going to be needed to get it moving. Q-branch has obviously tweaked the engine because the DB5 can certainly shift. And the faster it moves, the harder it is going to be to stop. At the last moment, Bond's high-tech car can't save him from a simple mirror trick that blinds Bond with his own headlights. With so much momentum behind it, it's no surprise that when it crashes, the car demolishes a brick wall.

At the end of it all Tilly Masterson is dead, as are a number of henchmen, and Bond is captured. It's not Bond's finest moment, but the Aston Martin DB5 became a fan favourite.[2] Its return to the franchise in 2012's *Skyfall* was a huge crowd-pleaser, even if it made no sense whatsoever.

## Throwing your weight around

Bad guys are always going to put obstacles in Bond's way. Watching him overcome, work around or simply smash through them is all part of the fun. Bond has mastered the art of demolishing things. Getting behind the wheel of some very heavyweight vehicles has certainly helped. His overweight DB5 could demolish a wall, but the crash also stopped the car in its tracks. The 42-tonne tank he borrows to escape an interrogation and hunt down the villains in *GoldenEye* just ploughs straight through a wall and keeps on going.[3]

At speeds of up to 35mph there isn't much that will stop a tank – not walls, statues or the St Petersburg traffic. Everything in its path is squashed, smashed or knocked out of the way. Yet despite this unmissable trail of destruction, the tank's huge size and low speed relative to every other vehicle on the road, Bond still manages to lose everyone and then sneak up on the villains without them noticing. Even more incredibly, no one is killed in

---

[2] More than 2 million toy versions were sold that came complete with spring-loaded ejector seat and a little plastic henchman to launch out of the car.

[3] Apparently, the tank needed a mile run-up to get to sufficient speed to break through the wall.

the entire process.[4] The only response from those clambering out of their wrecked car or dodging out of the way of the oncoming pile-up is bewilderment. Rather than mass panic, calls to emergency services and news cameras swarming all over the scene, bystanders seem to treat a Bond chase as a spectator sport.

While Bond villains try everything they can to stop Bond either heading towards or away from them, the rest of civilisation generally just keeps out of the way. Train passengers do nothing to alert the guards or driver to two people fighting on the roof or hanging off the side or the roof being ripped off the carriage with a digger. The train and the chase trundle on regardless.

Filming the car chase in *Diamonds Are Forever* through the streets of Las Vegas illustrates the point. Word had got out that the latest Bond film was on location on the city's iconic main strip. Crowds of onlookers filled the pavements and watched, unmoving, as stunt drivers sent cars hurtling round corners. There was nothing the filmmakers could do to disperse them and so they can clearly be seen in the background in the final film. Not that it matters, because everyone's attention is on the cars and the skill with which they are being driven.

Considering he is a secret agent, being conspicuous has never bothered Bond one bit. But I suppose when your life or the world is in danger you do whatever you can with whatever is available. If your car has been

---

[4] Deliberately showing civilians survive scenarios that should mean certain death (like being run over by a tank) was how the filmmakers could get a family-friendly certificate from the censors.

exploded into bits and there are people running after you firing guns, a canary-yellow, beaten-up, two-horsepower Citroën can seem like a good option simply because it is the only one available (*For Your Eyes Only*). When the gun-wielding goons pile into much faster Peugeot sedans to continue the chase, as long as the bad guys don't resort to any underhand tricks like mirrors or putting Bond's love interest bound and gagged in the middle of the road (*Casino Royale*), Bond has a chance. It is the skill of the driver that is all-important, and we know Bond has those skills in abundance.

A Bond car chase follows the general rules of Bond's world: the bad guys are methodical, almost conventional, but Bond gains the upper hand by improvising. The villains avoid crashing into anything but Bond, whereas Bond will deliberately slam into a sack of cement powder to create a dust cloud to mask his escape. In *For Your Eyes Only*, where the sedans initially stick to convention, driving along the roads, Bond takes a more direct path down the hill, through the olive trees. Piers, river banks, staircases, anywhere that appears to offer a clear-ish path is likely to be driven over. In *The Living Daylights*, when he runs out of road, Bond is happy to drive on a frozen lake.[5]

The chase starts when Bond's car is spotted as he and Kara Milovy, cellist, amateur assassin and Bond's love interest, try to escape across the Czech border to Austria.

---

[5] Due to a mild winter, the lake was not as frozen as the filmmakers would have liked and they nearly lost the Aston Martin through the ice.

Fortunately, he is driving what must be the most gadget-laden Q-car of them all. The police car that spots them is quickly dealt with using a laser beam in the hubcaps (see chapter 003). A couple of rockets clear a path through the lorry blocking Bond's progress. Bulletproof glass protects them from the large amount of ammunition fired towards them, and sled-skis emerge from the sides to help him drive on the snow and ice. Finally, a rocket motor helps accelerate the car over a dam and into a snow drift.

Gadgets or no gadgets, a car in a snow drift is next to useless. With plenty of buttons left unpressed and switches unflicked, and presumably yet more gadgets wired up to them that go unused, the car's final contribution to the chase is self-destructing in a huge explosion that knocks several pursuers off their skis. Wrecked cars and broken henchmen litter the path behind them, and the Austrian border isn't far away. But there are still snowmobiles and other skiers – all of them with guns firing – to continue the chase. Bond and Milovy have nothing but a Stradivarius cello to help them.

In Britain we are not often blessed with snow, so we are ill-prepared when the white stuff arrives. Therefore, when it does fall, in even the tiniest amounts, it is not unusual to see a Brit attempting to slide down any available slope on whatever improvised toboggan they have to hand – tea trays, plastic bags, anything really. A cello case seems almost practical by comparison.[6] You

---

[6] The British director of *The Living Daylights*, John Glenn, came up with the idea, but the screenwriter Richard Maibaum and the producers Michael G. Wilson and Cubby Broccoli (all snow-savvy Americans) took some convincing.

will also see these haphazard tobogganers failing to navigate their way down their chosen slope with any of the speed or grace shown by Bond and Milovy. While the rest of the world is laughing, British audiences are nodding and thinking 'I would' as the pair accelerate past the border checkpoint.

Gravity and the low friction provided by the compacted snow mean the cello case's trajectory will be downwards. Just why snow and ice are so slippery remains something of a scientific mystery. For a long time it was thought that skaters and skiers floated on a layer of water on the ice's surface. But water is considered a poor lubricant because it is not very viscous. A more recent study showed the layer of water also contains microscopic ice particles, which together form a fluid that can be as viscous as oil. The age-old practice of adding a hydrophobic wax coating to the underside of skis or snowboards repels the water and increases the viscosity further, reducing the friction even more. It would be sensible for Milovy to keep her valuable cello in a waterproof case, and it would also help when it needs to double as a toboggan. But they will need more than the cello stand as a rudder to guide them.

In reality the case was specially built out of fibreglass, with runners underneath and controls inside to help steer. Nevertheless, the pair regularly tipped over into the snow during the three days it took to shoot the sequence because Timothy Dalton is heavier than Maryam D'Abo and unbalanced them. But it's a fun and effective twist on the snow chase that features in so many of the Bond films. There's already plenty of variation in these wintry sequences, but the participants are usually being propelled along on equipment designed for the purpose.

Bond villains do like their winter sports – skiing, snowboarding, tobogganing, ice hockey and figure skating have all featured in the series. It started, of course, with Ian Fleming, who learned to ski when he was in Kitzbühel in Austria. When one of the local slopes was placed out of bounds because of the danger of an avalanche, Fleming – always fascinated by danger and still green to the sport – became obsessed. When he did pluck up the courage to try the forbidden slope, the reason for the warning became apparent. An avalanche started and Fleming was buried up to his shoulders. The rapid flow of tons of ice and snow can move rocks and uproot trees. Avalanches kill around 150 people every year through asphyxiation, exposure or physical trauma. Fleming was lucky to get away with nothing more than a few bruises, a twisted ankle and a great set-up for the fictional agent he would create years later.

Fleming's close encounter inspired the sequence he penned in *On Her Majesty's Secret Service*, where Blofeld tries to kill Bond by triggering an avalanche before he can escape his Alpine lair. Bond skis into a forested part of the slope so the snow is slowed by the trees, but not enough to kill several of the henchmen pursuing him. It also made an impressive moment in the film, as do all the snow- and ice-based chases.

Skiing and tobogganing take a certain amount of skill at the best of times – adding jumps, explosions and avalanches makes it much more difficult, but still too easy for a Bond chase. Combining all that and depriving Bond of either a ski pole or a ski is more of a test of our hero's great skills. And then the filmmakers have him ski over tables and into toboggan runs to make things really difficult. If you are impressed by what the ski-stand-ins and stunt team are doing, spare a thought for Willy Bogner, who directed many of these sequences and is doing almost everything you see on screen, but backwards and while holding a camera.

What they are doing really is dangerous. The toboggan/skis/bike chase in *For Your Eyes Only* went horribly wrong when the toboggan flipped and stunt actor Paulo Rigoni was trapped underneath. He was killed. It is the only death in the Bond films' 60-year history. So many of these amazing sequences have been created with so few incidents thanks to the immense talents of the team behind the Bond films and the safety precautions put in place.

Like the cello case mentioned earlier, vehicles and equipment can be adapted to the needs of the film and to reduce risks. They often have to be as heavily adapted as a Q-branch car. The Citroën 2CVs in *For Your Eyes Only* got an upgrade to 4CV. Extra handholds have been bolted on to trucks to give stunt actors something to hang on to. Modifications have been made to reduce the weight or improve manoeuvrability. A double-decker bus has had its top deck removed and replaced on rollers so it can be seen to be sliced off by a bridge (*Live and Let Die*). Sets are carefully prepared with ramps so drivers can put cars on two wheels (*Diamonds Are Forever*), and special rigs have been mounted on petrol tankers so they can be driven at 45°.[7]

---

[7] In the end the legendary Rémy Julienne, who devised and drove in so many cinematic vehicle chases, did the *Licence to Kill* stunt without the rig's help.

However well prepared and cleverly adapted the vehicles are, it is the skill of the drivers that makes the many chases so brilliant to watch. If you want an idea of just how good they are, take a look not at the carefully planned chases but at the moments they didn't expect and had to adapt to on the fly. The taxi chase in *Octopussy* has the classic bad guys running after Bond and Bond trying to escape. The vehicles speed through the packed street of Udaipur, past every Indian stereotype you can think of, while everyone on board tries to kill each other. At one point a Jeep full of henchmen pulls up alongside Bond's tuk-tuk to engage in some hand-to-hand and hand-to-five-bladed-dagger fighting. In the middle of all this the two vehicles separate to allow a cyclist, pedalling in the opposite direction, to pass between them, and the fight resumes once he has gone. This was unscripted. The guy on the bike didn't even know he was in a film; it was only the sharp reactions of the stunt actors that avoided a horrible accident.

While Bond's expertise on wheels, skis, tank tracks and horse have been proved, he doesn't always have a means of transport to convey him towards or away from the bad guys. Sometimes he doesn't even have an antique cello to help.

## Shanks's pony[8]
Audiences love the high-octane, fast-paced nature of the chases in the 007 films. You would have thought that

---

[8] Shanks are your legs, so to go by 'shanks's pony' means to travel around on foot.

having the protagonists simply running around would be too slow and too low-tech for the franchise's fans, but no one can be disappointed with the foot chase in *Casino Royale*. It again shows the skill and creativity of the people who make these films.

Bond is chasing the bomber Mollaka on foot. Bond is fast and agile, but he is outclassed by Mollaka at every step.[9] In order to keep up, Bond commandeers a digger to plough through the obstacles Mollaka is hiding behind. So Mollaka goes where the digger can't – up. But when he reaches the top of a crane, he still hasn't managed to shake Bond and, after a brief tussle, the only way to escape is back down. He jumps to a lower crane. The height of the drop and the small landing space make Bond pause, but there is no alternative and he makes the jump.[10]

There will be no give in the steel structure the two land on, so pretty much all of the impact will have to be absorbed by their bodies. The human body is remarkable for its resilience but it has its limits. The roughly 6m (20ft) jump means Bond, and Mollaka, will be travelling at about 40kph (25mph) when they land. This is usually the upper limit of what can be tolerated without severe injury. Despite the small surface area, falling on the feet is more survivable than falling on the head or side because the body has, to some extent, evolved to

---

[9] Mollaka is played by Sébastien Foucan, creator of freerunning and one of the founders of parkour.

[10] The jumps were, of course, done for real, only with safety lines attached to the actors in case they fell. The cranes were tied but still swayed a few metres in the wind.

withstand these kinds of impacts. One critical factor determining the amount of damage is how long a force is applied. By flexing the legs and crouching, the braking distance is increased. Even so, they will have to come to a complete stop within about half a metre.

Damage will occur at the weakest points; most commonly it is ankles and feet that are broken in feet-first falls. The tibia – the shin or shank bone – is the thinnest of the leg bones but even this can withstand a considerable impact. It is also supported by tendons and muscles. A certain amount of damage is to be expected, but adrenaline will get Bond and Mollaka through a lot of the pain and, as long as their legs aren't broken, they can continue the chase.

Bond lands the jump but only just. Mollaka lands better and takes another jump down onto a lower roof. This drop is much higher, around 11m (36ft), but there is more space available. Bond and Mollaka can disperse the greater energy they have accumulated in the fall by crouching and rolling onto their backs. People have survived falls from much greater heights, but very rarely without injury. Factors that improve the chances of survival are being in good physical shape, being relaxed, and training. This is where Bond's decades of experience of jumping and falling out of planes comes into play.

Bond seems free to run amok in any far-flung part of the world where a megalomaniac decides to set up shop. He can crash through construction sites, marketplaces and main streets while pursuing or being pursued by the bad guys, without worrying he'll be stopped by local law enforcement for so much as a speeding ticket. While it must cause endless headaches for M and Leiter, who have to apologise for the damage and smooth over the diplomatic fallout, audiences love it.

# 016

# *Licence to Kill* and a tanker full of cocaine

If you have ambitions for world domination, you are going to need funding. Volcano lairs don't come cheap. All those minions kitted out with custom boiler suits and matching machine guns are going to cost. Stealth boats come with large price tags attached. The sources of a Bond villain's apparently inexhaustible wealth vary, but they are almost exclusively illegal. Crime, certainly in 007's world, pays.

The international crime syndicate SPECTRE advertises one source of its revenue in its acronym: the last E stands for extortion. At the SPECTRE AGM, shown at the start of *Thunderball*, we get an insight into how they obtain the rest of their sizeable income. Each section head reports on the revenue they have obtained from blackmail, paid assassination work, consultation fees for major robberies, and narcotics. Smuggling drugs doesn't have the same associated glamour as that of gold, diamonds or jewels (see chapters 003, 007 and 012), but it does involve huge amounts of money, power and influence.

In terms of visual spectacle, drugs aren't in the same league as a giant space laser or nuclear missile holding world powers to ransom, but they really can have the power to exert control over governments. Several Bond films have referenced the illegal narcotics trade but both *Live and Let Die* and *Licence to Kill* made it central to the plot.

In *Live and Let Die*, Kananga plans to use heroin to make vast sums of money and shore up his dictatorship. Franz Sanchez, in *Licence to Kill*, uses the money from cocaine to control the government of a fictional South American country. Sanchez and Kananga have all the trappings of your typical Bond villain – luxury homes and spectacular lairs complete with well-appointed drug production facilities – proving the profitability of their ventures. How can these relatively simple chemical compounds exert so much power?

## Getting your reward

Humans, and other animals, don't always act in their own best interests. To ensure the survival of the species, our bodies and brains trick us into behaviour that helps us survive and procreate. For example, when we drink, eat seasoned food or engage in consensual sexual contact, our body releases chemicals called endorphins.[1] These endorphins lock into receptors in our brain, leading to the release of chemicals including dopamine as part of the so-called 'reward pathway'. The result is that we feel good, pain is dulled and the body becomes relaxed and sleepy. We are chemically encouraged to get the water, salt and sex we need to survive and thrive as an individual and a species.

Other chemicals – ones not normally found in the body such as cocaine and opiates – can also interact with this system of rewards to create the same effect. This has

---

[1] Endorphins is a compound word of endogenous (meaning within) and morphine because they produce similar effects on the body, though they are chemically very different.

pros and cons. Pain relief and sedation are invaluable in medicine. But the euphoria and feelings of well-being also produced by some of these substances mean they are open to abuse.

Any substance that can increase the effects of dopamine in the body has the potential to be addictive. Opiates can dock at opioid receptors, where endorphins normally sit, causing the release of more dopamine. Cocaine prevents the reuptake of dopamine and other chemicals important for mood, meaning they circulate for longer and increase the feel-good response.

The more the body is exposed to these external chemicals, the more it has to adapt to maintain normal function. In the example of opiates, more opioid receptors have to be made so that the body has enough to respond to the endorphins naturally present as well as the opiates. With more receptors, more opiates are needed to produce the same desirable effects, the baseline for feeling good is raised, and a tolerance begins to develop. Withholding these substances once a tolerance has developed will cause withdrawal symptoms because there aren't enough molecules around to trigger the body's normal feelings of well-being.

The impact on users can be dramatic. With opiates, their ability to sedate and relax the body can be pushed to extremes. It can mean respiratory collapse often combined with an inability to clear the throat of fluids, leading to asphyxiation. Excess cocaine, often from packets of the drug rupturing inside the bodies of those smuggling it, can lead to cardiac irregularities, raised blood pressure, stroke, haemorrhage and death.

When substances like cocaine and opiates are easy and cheap to produce on a huge scale, and your market has become dependent on them, staggering amounts of money can be made. An increased risk to health and well-being goes along with it. For the user, this can be from the effects of the drug itself or the circumstances surrounding drug addiction – through crime, violence, and disease transmission from sharing needles. With tens of thousands of deaths from overdose recorded every year, together with the huge sums of money and the networks of criminals and terrorists that exploit them, drug smuggling meets all the criteria you would expect for a Bond villain's criminal interests.

Bond films may pride themselves on cutting-edge technology and the latest science, but there is nothing new about the opiate trade. Dr Kananga's decision to deal in opiates, a trade that has been tried and tested for thousands of years, is a credible means of making a lot of money.

## Milk of paradise

Poppies were first cultivated around 5,000 years ago. Scoring the seed head, once the petals have fallen, releases a milky white latex known as the 'milk of paradise'. Left overnight, it solidifies and can be scraped off. This is raw opium. It has been used in medicine to relieve pain and, as many conditions cause pain, it has been formulated into countless remedies for practically every conceivable ailment.[2] Opium has no effect on

---

[2] Today opium is still part of the *British Pharmacopoeia* but the number of preparations has fallen from an all-time high of 38 in 1864 to just four in 1998.

the underlying cause of the symptoms, but the patient feels a lot better about them. Consequently, opium has been one of the few ancient medicines that actually benefited the people it was administered to.

Opium or the compounds within it, when administered to treat pain, do not automatically create addiction. It is prolonged exposure to the drug that does that. The enthusiasm with which opium has been doled out over the centuries to anyone and everyone, for anything and everything, undoubtedly led to addiction.[3] There have been many famous opium addicts over the centuries and while for some it released their creativity, it was clearly a debilitating and dangerous habit that was difficult to break. Demand for the drug has fuelled a sizeable trade for centuries, and even led to wars.[4]

Raw opium contains around 50 different alkaloid compounds, called opiates.[5] A handful have medical applications but only one is of any real interest to drug smugglers. That compound was first isolated by Friedrich Sertürner, a German chemist, in the early nineteenth

---

[3] Emperor Marcus Aurelius, who lived in the first century AD, is often cited as the earliest documented drug addict, though addiction wasn't really a concept in the ancient world. His symptoms suggest he suffered from withdrawal when the drug was withheld.

[4] The British bringing opium into China in return for tea, silk and other luxury goods led to the Opium Wars in the mid-nineteenth century. Opium had become a currency in its own right that could be exchanged for other goods, something smugglers can still take advantage of today.

[5] Compounds that produce opiate- or morphine-like reactions in the body, but aren't from poppies, are called opioids.

century. He first tested it on himself and some friends, causing them to fall asleep for 24 hours. He named the compound morphine, after Morpheus, the Greek god of dreams and son of Somnus, the god or personification of sleep (who gave his name to the opium poppy *Papaver somniferum*).[6] It took a while for the medical profession to catch on, but the benefits of a pure substance that could be given in controlled amounts, rather than impure raw opium, were clear. However, it also made addiction a lot easier, and things only got worse as the nineteenth century progressed.

The hypodermic syringe was invented in the late nineteenth century. Now, rather than smoking or ingesting impure opium, people could introduce pure morphine directly into the bloodstream, delivering it rapidly to the brain and its opiate receptors. Addiction rocketed.

Chemists searched for a way of modifying morphine, to offer the same or better pain-relieving qualities but without the dangerous addictive side-effects. Arthur Eichengrün and Felix Hoffman added two acetyl groups to morphine to produce diacetylmorphine in 1897, although who actually carried out the work is disputed.[7] The preparation was named 'diamorphine' and given the brand name 'heroin' because it was a drug for heroes.

---

[6] All varieties of poppy contain some morphine and related opiate compounds, but to varying degrees.

[7] Eichengrün was the senior chemist and Hoffman was a technician at the time. Augustus Matthiesson and Charles Romley Alder Wright, chemists working at St Bartholomew's Hospital in London, synthesised diacetylmorphine back in 1874 but it went unnoticed.

Diacetylmorphine is inactive in the body. What the acetyl groups do is increase the solubility of the molecule in fats, enabling it to cross the blood–brain barrier more easily than morphine. Once inside the brain, enzymes quickly remove the acetyl groups to convert diamorphine back into morphine where it can interact directly with the opioid receptors. Consequently, diamorphine acts more quickly than morphine and is far more potent – only a third of the normal morphine dose is needed to produce the same effects. The heroin high is higher, and the crash afterwards much worse, making it far more addictive. It was only a matter of time before heroin was banned in most countries around the world.

Morphine and heroin (called diamorphine so as not to alarm patients) still have medical uses, certainly in the UK, to manage pain. The ease of growing poppies, extracting opium, refining the morphine it contains and modifying this into heroin also makes it a very profitable street drug.

### Live and Let Die

The 1973 Bond film has Dr Kananga, dictator of the fictional island of San Monique, growing thousands of acres of poppies. He protects his plantations from nosy locals using camouflage nets and voodoo superstitions. The morphine extracted from these poppies is converted into heroin in his drugs lab, which he protects from prying eyes with a crocodile farm (see chapter 008). Distribution is carried out by his alter ego, Mr Big, through a chain of Fillet of Soul restaurants dotted throughout the US. The heroin, two tonnes of it, will initially be distributed for free, driving his competition out of business and, he expects, doubling the number of addicts or customers. Then of course he will start to charge for his product, at hugely inflated prices.[8]

The fictional Kananga has real-life equivalents, such as Frank Lucas, who was operating out of New York City in the early 1970s. At the time, the New York heroin business was run by the Italian Mafia, namely the Lucchese family, out of Harlem, with a seven-man African-American group called 'The Council' and a drug dealer known as 'Mr Untouchable'. Lucas began buying direct from manufacturers in Southeast Asia (not the West Indies, as in *Live and Let Die*), bypassing the Italian families and Mr Untouchable, to build a considerable business with links across the world. Law enforcement curtailed Lucas's global ambitions by arresting him in 1975.[9]

Kananga has much the same ambition as his real-life counterparts, but with the extravagant flourishes you would expect of a Bond baddie. His fictional drug lab, hidden within a ramshackle hut, seems very sophisticated. White walls and gleaming shelves stacked with bottles of brightly coloured liquids can be seen, along with a slick production line of white and brown powders making their way into packing boxes ready for distribution. The décor is certainly in keeping with Bond villains' preferred aesthetic of clean lines and shiny surfaces, but it looks,

---

[8] In Fleming's original novel, Mr Big makes his money from smuggling gold coins recovered from the wreck of a pirate ship.

[9] Mr Untouchable's name failed him in 1978 when he was also arrested.

like a lot of things in a Bond film, far more sophisticated than it need be.

The conversion process from morphine to heroin at a molecular level is complicated, but in practical terms it is so easy it can be done in a bathtub with a wood fire underneath. Of course, if you want to produce a high-purity product, a well-stocked lab certainly makes things easier, giving a higher level of control over the temperatures needed at different stages in the process and allowing monitoring of the pH to ensure acids are properly neutralised, but most drug smugglers aren't too bothered by quality control.

First, the raw opium is mixed with water, and calcium hydroxide (slaked lime) is added. The morphine separates and rises to the top so it can be scooped out for the next step in the process – adding ammonium chloride. This causes the morphine to become a solid that sinks to the bottom of the bathtub. It can be separated by filtering through muslin and is then spread out on baking trays to dry.

The morphine can then be converted into heroin. It is broken into small lumps (a hammer or cheese grater is all you need at this point) and then heated to drive off any water. Dumping the morphine into a big steel vat, or another bathtub, along with acetic anhydride followed by heating, causes a chemical reaction that adds the all-important acetyl groups to the morphine. When it is finished reacting, water is added to convert any leftover anhydride into acetic acid (the acid in vinegar). Impurities are removed with chloroform and the acetic acid neutralised with sodium bicarbonate (baking soda), a vigorous reaction that can cause the mixture to boil and explode. The resulting heroin can be filtered off from the mixture and dried for bagging up. The process is known as cooking, and requires nothing more than basic equipment, time and good ventilation.[10]

Dr Kananga's plan to dominate the US drugs market can't be allowed to go ahead. Bond sets fire to the fancy heroin lab and the scientists flee from their illegal work, presumably straight into the jaws of the waiting crocodiles. Then 007 jumps into a handy boat and escapes to reveal the extent of the plot to Felix Leiter. In spite of the fact that the drugs are headed for the US, and how much damage and destruction Bond achieves during his dramatic escape, the British agent is still trusted to go back to San Monique and destroy the poppy fields. He has Quarrel[11] with him, who helps set the charges in the poppy fields, while Bond goes to rescue Solitaire and kill Dr Kananga (see chapter 020). All of which he achieves, even though he is threatened by snakes, sharks and hordes of heavies. The poppies are destroyed and the day is saved once again. All the Americans had to do was provide transport, foot the bill and spend a lot of time apologising to the innocent people who have their lives and property disrupted by Bond.

If the wholesale destruction of Kananga and his heroin operation is meant to deter others from taking up the

---

[10] Some steps can take hours and give off very smelly gases. The strong smell of urine generated when the ammonium chloride is added can easily give away the location of your heroin production plant to law enforcement.

[11] Son of the Quarrel who was burned to death in Dr No (see chapter 022).

trade, Bond and the CIA have learned nothing from history. The major criminals in 007's world may have decided to focus on other evil plots for a few years, but drug smuggling is too lucrative to ignore. A few Bond films later and the villain, Kristatos, in *For Your Eyes Only*, is smuggling raw opium hidden inside giant rolls of newsprint paper.[12] A stray bullet in a gun battle pierces the casing and Bond is able to identify the brown sticky substance leaking out of the rolls by the characteristic bitter taste.

In *The Living Daylights*, as part of a convoluted plan to enrich themselves, the bad guys are involved in arms dealing and opium smuggling out of Afghanistan.[13] But the drug smuggling endeavours in both these films are minor sidelines to the main plots. The following film, *Licence to Kill*, made drugs central to the bad guy's plans. Again it is an American issue, but again it is the British agent who saves the day. This time, however, Bond is acting on his own initiative and outside of M's authority.

### Licence to Kill
The opening of the 16th 007 instalment has Timothy Dalton's James Bond invited along to thwart a drug smuggling operation. He helps capture the plane carrying

---

[12] The smuggling set-up taken from Ian Fleming's short story 'Risico'.

[13] The film was released in 1987 and the plot shares many parallels with the Iran–Contra affair that was going on at the time. Lieutenant Colonel Oliver North facilitated the illegal sale of weapons to Iran and siphoned off the profits to help fund the Contras, Nicaraguan rebels fighting against the Sandista rulers.

the drugs with a daredevil stunt even though he was supposed to be going to a wedding – he just can't help himself. Bond's friend and CIA contact Felix Leiter is just as guilty of never being able to switch off – it's his wedding and he is the one who is tempted away from his big day to catch baddies, taking his best man with him. With the smugglers caught, Bond and Leiter parachute down in front of the church before the bride has had a chance to reconsider her life choices.

The wedding goes ahead but married bliss is short-lived for Mr and Mrs Leiter. The drug smugglers turn up to take their revenge before the couple can leave for their honeymoon. Mrs Leiter is killed and Felix is almost killed by a shark (see chapter 008). Bond sets off to avenge Mrs Leiter's death and Felix's attack, going against M's express orders not to interfere in something that is none of his business, like that has ever stopped him before.

Once again we have a criminal mastermind hell-bent on making huge amounts of money from the industrial-scale distribution of illegal drugs, and using the power and money that come with it to gain influence over the governance of certain countries. In this case the villain is Franz Sanchez and the fictional country is the South American Republic of Isthmus. And the drug is cocaine, not heroin.

Cocaine is the main active chemical in the leaves of *Erythroxlon coca*, a shrub that grows in South America and Southeast Asia. When sniffed, smoked or injected, cocaine rapidly penetrates the brain. The increased circulation of dopamine this causes gives a feeling of confidence, optimism, boundless energy and increased self-esteem.

Indigenous tribes in South America have been happily chewing coca leaves for at least 3,000 years to reduce fatigue and suppress hunger, incorporating it in religious festivals and making use of its analgesic properties.[14]

Pure cocaine was first isolated in 1860 by German chemist Albert Nieman. When he tasted his new compound he noticed his tongue went numb. Cocaine's usefulness as a local anaesthetic wasn't fully realised by Europeans and put into practice until the 1880s.[15] But the beneficial qualities also come with less enjoyable side-effects, particularly when taken in larger amounts. They include slurred speech, tremors, chest pains, hallucinations (a sense of insects crawling on the skin) and paranoia. Cocaine contracts blood vessels and can restrict blood supply, damaging parts of the body, and causing chest pains, asthma and collapse of nasal tissues. Medicinal use of cocaine has declined considerably, but enthusiasm for its recreational use continues unabated. And as long as there is demand, there will be suppliers.

Coca plants live for up to 30 years and can produce five crops a year. The initial processing to extract the tiny amount of cocaine present in the leaves can be done by the farmers. First, the leaves are dusted with cement before being chopped up into little pieces using a strimmer. The pieces are then soaked in gasoline to extract the cocaine. Sulphuric acid is then poured into the gasoline and the

---

[14] It is also an important part of certain rites and festivals.

[15] Most local anaesthetics used today are derivatives of cocaine. Novocaine (or procaine) was introduced in 1905 as a less toxic synthetic alternative. It doesn't produce the same stimulatory effects and therefore is not subject to the same levels of abuse as its parent compound, cocaine.

cocaine moves into the watery acid layer. This is siphoned off and caustic soda is added to turn the cocaine into a solid 'pasta', meaning 'paste'. The pasta is separated by filtering through cloth, forming an easily transportable lump containing 40–60 per cent cocaine. There is no other crop that can provide such a reliable and profitable income, but coca farmers are not rich.

Drug smugglers buy the pasta from the farmers and take it to big industrial labs hidden deep in the jungle, like the one we see in *Licence to Kill* but bigger in scale and without the fancy exterior.[16] Two more purification steps are carried out to produce pure cocaine hydrochloride, ready for shipping to customers. Each refinement increases the value of the product, but taking it across the border into the United States is what really inflates the price and the profits.

Like his real-life counterparts, Sanchez is using every available method to smuggle his cocaine into the United States. Private planes, like the one intercepted by Bond and Leiter at the start of the film, are pretty standard. Hiding the bags of cocaine in the bait and sand that are part of an exotic fish emporium is a little more unusual.[17] To scale up his operation, Sanchez plans to use petrol tankers, with the cocaine dissolved in the petrol to avoid detection. The filmmakers based this on reports of how smugglers really do use petrol tankers to smuggle cocaine, but with a few Bondian embellishments.

---

[16] Exterior shots for Sanchez's lair were filmed at the Otomi Cultural Centre in Toluca, Mexico.

[17] An idea taken from Fleming's *Live and Let Die*, where gold coins were smuggled in the sand at the bottom of fish tanks.

On a tour of his jungle lab facilities, Sanchez explains to his prospective clients that the cocaine will be delivered to them by tanker, but it will be dissolved in the petrol within the tanker to avoid detection. The tanker comes with a chemist to help them separate the cocaine when it arrives. They get to keep the petrol as a bonus. It is implied that the separation process is difficult and that Sanchez wants to protect his intellectual property to retain his customers. However, as we have seen, that can be done easily with a few household chemicals.

Transporting cocaine in petrol tankers is still a risk, especially when those tankers are being pursued by James Bond. Sanchez, all the cocaine and his whole operation are brought to an explosive and fitting end, ignited with the lighter Felix gave Bond as a gift.

Cocaine really has infiltrated and influenced many governments, and even financed one *coup d'état*. Some traffickers, like Carlos Lehder, have bought islands in the Bahamas, a popular haunt of Bond villains over the years, and made them their base for wild parties and an endless stream of cocaine arriving from South America before it is flown on to the United States. Norman Cay, Lehder's home for a while, even had sharks swimming in the bay.

Other drug lords have accumulated so much wealth they have offered to pay off their nation's debts in return for being left to their illegal activities unmolested. Pablo Escobar employed an army of 3,000 and equipped them with some serious military hardware including ground-to-air missiles, and came close to buying 120 stinger missiles. It makes Sanchez's fictional operation, with no more than a few dozen goons and only a couple of stinger missiles at his disposal, seem low-key and ill-equipped.

The fictional Sanchez is most often compared with the very real Manuel Antonio Noriega Moreno, de facto ruler of Panama in the late 1980s. He amassed a huge personal fortune through the illegal trade of cocaine, among other profitable crimes, and built up an effective dictatorship. The deteriorating relationship between Panama and the United States prompted an invasion. Noriega was eventually captured, on 3 January 1990,[18] six months after the release of *Licence to Kill,* and taken to the US for trial. Sanchez's fictional exploitation of cocaine certainly mimics real life but, for once, the fantastical fictional world of 007 is but a pale imitation of the excesses that real drug traffickers have gone to.

---

[18] He had sought sanctuary in a church and was finally prised out by playing pop music at full blast 24 hours a day. The Clash's 'I Fought the Law' was apparently the final straw that broke either Noriega or the priests.

# 017

# *GoldenEye* and the EM pulse

Weapons are a common feature of 007 films, but some of those weapons are very uncommon. There are all the standard items you would expect a secret agent, villain or henchman to arm themselves with, such as guns, bombs and knives. Some of these have been specialised to make them more interesting, such as plating them in gold (chapter 009), salting them with a few extras like cobalt (chapter 013), or adding poison to their tips (chapter 002). These conventional weapons have also been adapted to fit into unexpected places such as inside cigarette cases, under a fake meringue, or into a shoe.

There are lots of other weapons you might not expect to see in spy films, for example harpoons, bows and arrows, and swords. They may not be as common, but there is no mystery as to what they are or how they might kill. Even invented weapons – like the yo-yo saw in *Octopussy*, and its grandiose descendant the five circular saws hanging from a helicopter in *The World Is Not Enough* – don't need explanations as to why you should avoid encountering them. Sharp edges and pointy things can cut into or pierce vital organs and blood vessels whether they are at the end of a knife, harpoon or airborne circular saw.

In the same way improvised weapons, like the nail gun in *Casino Royale* or surgical scalpels in *Diamonds Are Forever*, don't need any explanation. But there are other weapons in Bond's world where it is not immediately obvious what they are or how they could cause harm. Usually they come with an explanation, but sometimes the audience simply has to accept that what they see on screen can kill or harm a person. These unusual weapons are not always the invention of the filmmakers either – some are based on solid science.

The biggest and baddest of all the weapons in the 'unusual' category has to be the EM pulse bomb named GoldenEye.[1] It is so impressive it gets to be the title of the film and enjoys a detailed explanation of how it works from none other than Dame Judi Dench herself. The weapon, a nuclear bomb, is fairly conventional, but the intention of this weapon is not what you might expect – the destruction of not buildings and people but electronics.

The grand plan in the 1995 film *GoldenEye* is to explode a nuclear device in the air above London. The pulse of energy that is released will knock out electrical systems and wipe computer records. Criminal convictions will be erased, along with illegal money transfers, and send England's capital into chaos. But would it, or any of the other weird and wonderful weapons in the franchise, really work?

### Start as you mean to go on

The trend for unusual weapons started, of course, with Ian Fleming's novels. Fleming, always a fan of gadgetry,

---

[1] The name was taken from Ian Fleming's Jamaican residence. Fleming named his house after the Second World War operation Golden Eye that he was involved in. He helped put together plans to maintain an undercover team of operatives in Gibraltar should Spain be invaded by Hitler's forces.

guns and the latest technology, incorporated as much as he could into his stories. He also weaponised animals (see chapter 008), plants (see chapter 006) and medical equipment. The Hercules Motorized Traction Table in the novel *Thunderball* is definitely in the unusual category.

While Bond is having his tight muscles stretched out, a sinister hand appears in his line of sight and turns the pressure dial up, past 150lb (as shown on screen – equivalent to 68kg) into the red numbers and leaves the needle hovering at 200lb (90kg) of pressure. Apparently this could have been enough to kill Bond, though how I'm not sure.[2] In centuries gone by, people have been tortured by being stretched on the rack. Muscles could be torn, ligaments snapped and bones dislocated but the victim had to survive so they could divulge information. Fortunately, the machine in *Thunderball* is switched off before any major damage can be done and Bond is given a shot of coramine,[3] a stimulant, to improve his pulse, which for some inexplicable reason isn't already racing from the stress of the situation.

For the benefit of the big screen, some aspects of Fleming's imagination had to be reined in, while others provided the mere seeds that allowed the scriptwriters to grow them into fantastical set pieces and plots. The assassination by traction table was deemed within acceptable bounds, and features in the film version of

---

[2] And if it was enough to kill, why would you build a machine that powerful in the first place?

[3] Now known as Nikethamide, it is sometimes used in sports doping. In the mid-twentieth century it was used as an antidote for overdoses of tranquillisers.

*Thunderball.* However, when it comes to other unusual weaponry, Fleming was just the beginning.

Take *Goldfinger* as an illustration. The filmmakers updated the weapon threatening to cut Bond in half, from a circular saw to a laser (see chapter 003), and they gave Goldfinger a dirty bomb to play with (chapter 008). On the other hand, they retained Oddjob's hat, a bowler with a rim reinforced with a 'light but very strong alloy' that could have 'smashed a man's skull or half severed his neck'. Only minor changes were needed from page to screen. Instead of throwing the hat at the wood panelling in Goldfinger's home, as Oddjob does in the book, he uses the bowler to decapitate a stone statue. The visual demonstration is perfect and the film doesn't need to slow down for lengthy descriptions or explanations. Later, when the hat is thrown towards Tilly Masterson's neck, we know that it has killed her. Some of the other unusual weapons in the series don't even get an introduction before being used to bump off characters.

**What was that?**
Five films after *Goldfinger*, and audiences watching 1973's *Live and Let Die* weren't going to be fazed by a few unusual weapons appearing on screen. When a man is shown wearing an earpiece, followed by shots of a strange hand swapping some connections, all that is needed is a whirring sound and a visual of the man collapsed on his desk to understand he has been killed. No explanation is offered, but it is clear his death was caused by something deliberately sent through his earpiece, and not natural causes. The implication is that

the man has been killed by sound, but the film doesn't give the audience time to think about it and moves on to the next scene before anyone can start asking awkward questions.

It may sound silly but sound really can be a weapon. There are two main factors that need to be considered: frequency and decibels. Frequency is the pitch of the noise (measured in Hertz (Hz), or oscillations per second), and decibels measure how loud the noise is. Another important consideration is natural frequency. Everything has a natural frequency. If an external force is applied to an object at its natural frequency it will oscillate. More force, or more decibels, and that oscillation will be greater. For example, when a pendulum is released it oscillates back and forth at its natural frequency. Push the pendulum harder and it will swing further. If a force continues to be applied to an object at its natural frequency, they reinforce each other and the movements are amplified. In terms of hearing this is a good thing: different frequencies trigger the movement of different tiny hair-like structures within the ear that send signals to the brain where they are interpreted as sound.

Humans can hear frequencies in the range between 20 and 20,000Hz,[4] but other frequencies can still have an effect on our body, even when we can't hear them. Just below the range of human hearing at 19Hz is the natural frequency of the fluid in your eyes. This 'sound' can therefore make your eyeballs wobble. Frequencies between 0.5 and 0.8Hz can interfere with your breathing

---

[4] The exact range obviously varies between individuals and changes as we age.

and shake your bones. The poor victim in *Live and Let Die* may have been on the receiving end of a loud blast of a sound at a frequency that shakes blood vessels or brain matter.[5] The vibration the noise created could have caused bleeding in the brain. This can have very sudden and dramatic consequences because the brain is housed inside an essentially closed space. The build-up of pressure as more blood flows from the body into the limited space inside the head can cause more damage to nerves and tissue. Collapse and death can be rapid.

The sound weapon in *Live and Let Die* is an excellent choice, tailored as it is to the victim and the situation – a UN conference where the target, a British ambassador, is likely to wear a convenient earpiece for listening to the translation of the speeches being given. The controls for the earpiece are also in a separate location where few people will be paying attention if a wire or two are swapped around. Villains in other Bond films have been equally meticulous in tailoring their unusual weapons to the situation, but on a much grander scale.

## Think big

*A View to a Kill* from 1985 features a very conventional weapon used in a very unconventional way. The villain, Max Zorin, plans to use explosives to trigger an earthquake.[6] Zorin's motivation is greed. He has a

---

[5] I'm imagining the diplomat's brain wobbling around like the aliens' brains in *Mars Attacks* when they are subjected to Perry Como's singing.

[6] A similar idea was used in 1980's *Superman II* film, where Lex Luther plans to sink a large chunk of California into the ocean to improve the value of his inland real-estate.

successful microchip manufacturing company, with several government contracts, but he wants a worldwide monopoly and doesn't care that thousands will die in his plan to wipe out the competition – the companies based in California's Silicon Valley. Their location is to Zorin's advantage, as the valley is situated between two fault lines, the San Andreas and the Hayward. These cracks in the Earth's crust are at the boundary between two of the enormous tectonic plates that cover the surface of our planet like a jigsaw puzzle. These tectonic plates are slowly being pushed past each other by the convection currents generated in the hot, viscous mantle beneath. The result is earthquakes. Sudden shifts of huge amounts of rock cause cracks to appear and sections to crumble away. Zorin is just two earthquakes away from sinking Silicon Valley, and his competitors, into oblivion.

Earthquakes are guaranteed, but when and where they will happen is almost impossible to predict. Zorin has no time for nature to take its inevitable course, so he decides to help things along. His plan is two-fold. First, sea water is pumped into Zorin's oil wells to put pressure on the Hayward fault. This is a well-known potential problem in mining and drilling for oil. There have been many recorded incidents where human activity underground has triggered seismic activity.

In 2007 miners were working 450m (1,476ft) underground to extract coal from the Crandall Canyon Mine in Utah. In the early hours of 6 August the pillars supporting a section of the mine gave way and 200,000m² (2,152,780ft²) of earth collapsed, trapping the miners. The owner of the mine said an earthquake measuring magnitude 3.9 was the cause of the collapse. However, further investigations revealed that it was the other way around. It was the collapse that had triggered the earthquake. Seismic activity as a result of humans extracting petroleum resources has been recognised since the 1920s. Zorin's plan to inject water into his mine has potential consequences, particularly in an area as geologically volatile as the Hayward fault, though he may not get everything he wants.

The second part of the plan is to create a massive explosion in a cavern under the San Andreas Lakes that will trigger another earthquake along the San Andreas fault. The explosion will break through the lakebed, flooding the fault below. Water is heavy, and the movement of large quantities of it can trigger earthquakes, as has been demonstrated when building dams.[7]

When making *A View to a Kill*, the filmmakers went to considerable trouble to get their science right. The film's co-writer, Michael G. Wilson,[8] said, 'This is something that could almost be done. In other words, once you see this picture, you'll see that it's a little frightening. Everything from a geological standpoint is absolutely true.'

Zorin's plan to trigger an earthquake is certainly scientifically and geologically plausible. TNT is a powerful explosive, but the energy needed to shift

---

[7] A number of earthquakes were triggered in the Lake Mead area in the 1930s as water accumulated behind the newly constructed Hoover Dam.

[8] Stepson of Bond producer Cubby Broccoli. His background is in electronic engineering and law, but he has gone on to not only co-write 007 screenplays but also produce the films with his stepsister Barbara Broccoli.

tectonic plates is on an entirely different scale.[9] Nuclear
bombs have triggered earthquakes, and aftershock
sequences, but they were less powerful than the initial
explosion. If you want to create economic havoc by
destroying a major industry, and you have no moral
qualms about killing tens of thousands of people in the
process, then Zorin's plan might work. But there is no
guarantee it will.[10] He also needs considerable financial
outlay to buy up the oil wells and mines, and staff to fill
it with tons of strategically placed TNT and fertiliser, but
I'm sure Zorin has done the cost–benefit analysis ahead
of time.

Most Bond villains' big plans seem to operate on
narrow profit margins, though I guess you can't put a
price on power or satisfying your ego. Zorin may have
one financial advantage, as his silicon chips aren't just any
old silicon chips. He can command a premium, not just
as the only remaining major manufacturer, but because
his chips are resistant to electromagnetic pulses. The
importance of this resistance is illustrated in a later Bond
film, 1995's *GoldenEye*.

## Lights out

The bad guy in *GoldenEye*, former British secret service
agent 006 Alec Trevelyan, plans to exploit a vulnerability
within electronic devices to steal millions of pounds and
wipe all evidence of his crime, as well as any other

---

[9] How silly would Zorin look if he went to all that expense and
effort only to produce a minor geological wobble.

[10] Although an earlier draft of the script apparently had Zorin
diverting Halley's Comet to crash into Silicon Valley, making this
plan look considerably more realistic by comparison.

electronic trace of his existence. We're not talking about a simple power cut. It takes more than pulling out a few plugs and cutting one or two wires to achieve wholesale obliteration of electronic records. What Trevelyan needs is a massive pulse of energy to surge through all the wires and circuitry in London to destroy or corrupt everything.

Sudden surges in the flow of electricity can overload wires and electronic components, causing them to fail. It's why fuses are fitted to devices, so the fuse will break, preventing further damage to the device it is wired into. But these surges can come from elsewhere, not just the mains supply.

Nature produces electromagnetic pulses (EMP) that can disrupt the normal function of electronic devices. The largest source of EMPs is the sun and the solar flares that erupt from its surface. The sun has complex and constantly changing magnetic fields that sometimes become tangled releasing vast amounts of energy, charged particles and radio waves into space. If a solar flare happens to explode in the direction of the Earth then the charged particles it has spewed out can show up here as the bright lights of the Aurora Borealis and Aurora Australis. And the radio waves that go along with the flare can disrupt long-range radio communications. We are protected from the worst of these effects by the Earth's own magnetic field, but satellites and astronauts in orbit are much more vulnerable and sometimes have to take extra precautions.

Though solar flares release staggering amounts of energy, they occur far away. Smaller releases of energy can be more damaging if they occur nearby. An example would be another naturally occurring source of EMPs, lightning. These bolts of electrical energy are very powerful – a single lightning bolt can be tens of thousands of volts – but the damage is generally very localised. Our modern-day reliance on a constant supply of electricity means blackouts caused by lightning strikes can be very disruptive and data can be lost if it is not backed up. Damage to electronic equipment used to fly planes or control trains could be lethal. It is a weak point that could potentially be exploited and so research has been carried out into how to protect critical systems.

To develop protective measures against huge EMPs, you must generate them so their effects can be studied. In the 1970s, at a remote site in New Mexico, a landing stage was built almost entirely out of wood onto which planes would be wheeled and subjected to intense beams of electromagnetic radiation.[11] It's called 'the trestle' and, though it is still standing, it's no longer in operation. The main problem the researchers faced was simply the amount of energy needed to generate a big enough EMP to affect an aircraft. Aircraft are one thing, but to produce a pulse that could affect something as big as a city, you need a nuclear weapon.

The electronic side-effects of detonating an atomic device were discovered by accident during a nuclear explosion in 1962, not 1945 at Hiroshima as stated in the film *GoldenEye*. In a test codenamed Operation Starfish Prime, a thermonuclear device (see chapter 013)

---

[11] The photos I have seen honestly look like what a Bond villain would build if that Bond villain went in for shabby chic.

was detonated 400km (250 miles) above the Earth's surface over the Pacific Ocean. It generated a pulse of electromagnetic radiation that created an aurora visible from both sides of the equator that lasted for 15 minutes. It knocked out electric lights 1,280km (800 miles) away in Hawaii and disrupted satellites orbiting above the region of the explosion. It sounds impressive, but it is not nearly enough for Trevelyan's plans.

Not everything went dark in Hawaii; only 30 strings of streetlights (1 per cent of the total), already in a poor state of repair and particularly vulnerable to the pulse generated by the bomb, were affected. The damaged satellites continued to work, just not as well.[12] That large amounts of energy were released during a nuclear explosion was not a surprise, but the effect it had on electronic devices, and how distant those effects were felt, was.

The idea that atomic weapons could be specifically exploited for their effects on electronic systems, rather than general mass destruction, was an interesting military prospect. A weapon that could cause not just a blackout but damage electronic equipment to the point where data is lost or corrupted has many advantages. There is no direct loss of life or damage to property and no lingering harmful radiation or chemicals.

In 1997 Major Scott W. Merkle wrote a paper for *Military Intelligence* magazine that stated EMP weapons

---

[12] Their solar cells were affected by electrons released in the explosion that became trapped by the Earth's magnetic field. The solar panels struggled to generate sufficient charge to power the satellites.

had been built but hidden from the public by a fearful government. According to his paper, a declassified military report claimed a one megaton bomb detonated 1,280km (800 miles) over Nebraska would disable every computer in the United States, southern Canada and northern Mexico. His article offers only one reference to back up his claim, an article from 1995 published in the Canadian paper *The Globe and Mail*, but I can find no trace of it.[13]

Operation Starfish Prime and the testing carried out at the trestle shows why, in the film *GoldenEye*, Trevelyan can't just build his own EMP weapon out of a few bits of spare wire and some batteries.[14] He must go to the considerable trouble of stealing an existing weapon from the Russians.

The film's titular GoldenEye system consists of two nuclear devices in orbit. Trevelyan's team explodes the first over a Russian military base that destroys the base and two jet fighters and causes an electricity blackout for 48km (30 miles) in every direction, as well as knocking out a few satellites that happened to be orbiting in the general vicinity. It is an effective demonstration of the weapon's potential, with everything even vaguely electronic exploding and blue sparks and electrical

---

[13] If you have a copy of this article I would love to see it.

[14] But apparently Q can. In *No Time to Die* he has managed to shrink an EMP device to a size that fits inside Bond's watch. It is powerful enough to knock out CCTV cameras and henchman Cyclops's bionic eye, killing him. It doesn't seem to affect Bond's electronic earpierce, however. Q must have built it with EMP-resistant chips, *à la* Zorin.

flashes appearing all over the place followed by complete darkness.

The confusion created by the first weapon provides cover for the theft of the GoldenEye guidance system. Trevelyan gains full control over the second nuclear device and no longer has to traipse all the way to Siberia to detonate it. He can make his illegal financial transactions by hacking into London banks, trigger the second GoldenEye weapon to cover his tracks, and all from the comfort of his secret lair.

One thing he does need is a satellite dish to transmit his instructions to the orbiting nuclear weapon so it will detonate in the right place at the right time. What he doesn't need is such a big satellite dish, other than to match the size of his ego and provide a very cool location for the film. The Arecibo radio telescope in Puerto Rico was, at the time of filming, the largest of its kind in the world. At 305m (100ft) in diameter, it was an impressive place to film Bond's final confrontation with Trevelyan and where the bad guy meets a fitting end, impaled by the antenna of his own satellite dish.

# 018

# *Tomorrow Never Dies* and the stealth boat

Bond and his enemies always travel in style. For professional reasons you would think that few people in 007's world would want to attract attention to themselves, but their chosen means of transportation can rarely be described as discreet. Bond villains don't usually go out of their way to blend in with the crowd, but given their general contempt for the crowd, that is hardly surprising. And if Bond needs to move in such antisocial circles, he needs the appropriate vehicle to do it in.

Whether it's a private jet, an armour-plated train or an expensive car, the exterior will be impressive and the interior will be plush and comfortable, and it will come with all sorts of extras. No expense is spared, and there is no compromise when it comes to practicality, though what is practical in Bond's world is a little different to what is practical in ours. Fuel efficiency or depreciation are not important considerations. Bond villains always have plenty of fuel sitting around the place and few of their, or 007's, vehicles will survive long enough to be traded in for newer models. The must-haves for these vehicles – bulletproof glass, rockets or stealth technology – are not the sorts of things you see listed as optional extras even in so-called supercars.

The Bond franchise contains an eclectic mix of classic cars and innovative vehicles. Many will be cleverly adapted to suit the strange situations international spies and criminal masterminds sometimes find themselves in. Some are surprisingly versatile. Cars can double as planes or submarines and, to complete the circle, an aircraft doubles as a submarine in *No Time to Die*. Most vehicles in the real world stick to performing well in just one medium, be it air, water or land. Sure, there are amphibious vehicles, but none of them can submerge completely, or look as good as Bond's Lotus in *The Spy Who Loved Me*. In this fictional world, style is every bit as important as substance.

Elliot Carver, in *Tomorrow Never Dies*, is an exception in the Bond franchise, and not just because he is an unscarred British bad guy who doesn't even have a white cat or a shark pool to advertise his evilness. While most master criminals like to live their lives in well-appointed seclusion, Carver is a very public figure, head of a global media corporation. His fellow megalomaniacs travel about in the kinds of planes, trains and automobiles that get noticed. However, the point of Carver's catamaran is that it can literally slip under the radar.

### Under cover

It makes sense to disguise what you are up to if what you are up to is no good. There are many ways to do this. Dr No starts rumours of a dragon infesting his swampy island lair to keep the locals and troublesome secret agents away. In Fleming's novel, the dragon is an enormous vehicle with tyres taller than Bond, wheel arches extended into wings, an extra trailing wheel for stability that has been made into a tail, and a metal snout added to the radiator complete with flamethrower.[1] It

---

[1] The vehicle was based on a swamp buggy he and his friend toured around in on their visit to the island of Inagua (see chapter 001).

might not convince as a living, fire-breathing dragon, especially with the sound of a diesel engine emerging from it, but it would certainly be intimidating.

The dragon in the film version of *Dr No* perhaps marks the point where the tiny, stretched budget finally snapped. With more cash to back up Ken Adam's creativity, the dragon tank could have been a magnificent mechanical creature rather than a boxy, painted, plywood-covered truck. The flamethrower is impressive, but the idea that either Quarrel or Ryder would believe it was an actual dragon is an insult to everyone's intelligence. In a similar vein is the iceberg submarine from *The Spy Who Loved Me* and the one-man crocodile sub from *Octopussy*. As far as disguises go, they are the vehicle equivalents of fake moustaches.

The revolving number plate on the Aston Martin DB5 in *Goldfinger* was a low-tech but slightly more credible way of avoiding detection, or at least a few parking tickets.[2] As stealth technology goes, a revolving number plate is pretty basic, but everyone has to start somewhere. The first person in the franchise to take stealth seriously was Hugo Drax in *Moonraker*. His ability to hide an enormous space station from absolutely everyone stretched credibility to breaking point (see chapter 005). More realistic is Carver's stealth boat in *Tomorrow Never Dies*.

In the 1997 Bond film, Elliot Carver, owner of the Carver News Corporation, has just opened a new office

---

[2] This is apparently why the gadget was included in the film, even though Bond doesn't use it on his mission. One of the filmmakers had recently received several parking fines and wished out loud that he could change his number plate to avoid them.

in Hong Kong with plans to expand into China.[3]
Having established a toehold in the Asia market, he
plans to increase his slice by getting the scoop on all
the other news agencies. Rather than wait for the news
to happen, he decides to make sure it occurs when it
suits his news bulletins and print deadlines. With
cameras ready to film the fallout, Carver tries to
provoke a war between China and Britain by firing
stolen British missiles at Chinese ships.

Carver's declaration 'You give me the pictures. I'll give
you the war' are words that have been attributed to
William Randolph Hearst. Hearst was talking about the
US war with Spain in 1898, which he hoped to use to
increase sales of his newspapers, not by reporting the
facts but by deliberately stoking anti-Spanish sentiment
among Americans via sensationalising events or just
making them up.

Back in Bond's world, for Carver's plan to work he
can't be seen to be the one launching the missiles. All
the blame has to fall on the British and Chinese
militaries, and Carver's own ship has to remain hidden,
hence his state-of-the-art stealth ship. Though
modelled on the real-life US Navy's *Sea Shadow*,[4]
with 12 bunks and a basic kitchen (a microwave, fridge
and not much else), it is not quite up to a Bond
villain's high standards.

---

[3] Rupert Murdoch acquired the Hong Kong company Star TV in
1993 and set up offices all over Asia, making his company News
International one of the biggest news agencies in the East. His plans
to broadcast in China were halted by the Chinese government,
who effectively blocked transmissions into the mainland.

[4] A prototype was built but the ship was never commissioned.

The basic look of the *Sea Shadow* is retained for the film because it is key. Not only do the straight lines and strange angles appeal to the Bond villain aesthetic, but they are necessary to deflect radar signals and avoid detection. So that the radio waves aren't reflected back to the radar detector, the bows of stealth ships are angled inwards, curves are squared off and features that might be used to identify a vessel are boxed in to distort them. To make things even more difficult for the enemy, materials such as carbon fibre can be used that are transparent to radio waves, so the locating signal goes straight through the outer skin of the ship and is reflected off internal structures. With careful planning, these internal structures can be designed so that any reflected rays reveal a confusing or much smaller shape. This will certainly be needed for Carver's craft because it is very big. It needs to fit in all those minions, stolen missiles, a big sea drill and still leave space for Jonathan Pryce to stalk around and chew up the high-tech scenery.

The problem is that carbon fibre is very stiff and brittle and would break under the stresses and strains experienced by larger vessels in an ever-shifting sea. It's easier to build big boats out of cheaper, stronger materials such as steel to stop it being broken apart, but steel is very good at reflecting radio waves.[5] Special coatings can be applied to the exterior that absorb radio waves and transform them into heat, which is then absorbed by the metal structure it is painted on. It's expensive (not that

---

[5] Composite materials can also be used to strengthen the carbon fibre but may increase its detectability.

expense has ever bothered a Bond villain) but effective. However, it does present another problem.

Radar isn't the only way to find a vehicle. The engines powering them generate heat, which can be detected by infra-red sensors. Generating more heat from anti-radar coatings would only make things worse. Apparently Carver's craft avoids the problem of heat by running off two all-electric engines, which have the added bonus of being very quiet so submarines will have greater trouble listening for tell-tale engine noise. Carver can also bellow his orders to his minions more effectively. But again, with each new solution comes a new problem.

All-electric ships are certainly available today, but their range is limited and there are no recharging stations in international waters. As usual in a Bond film, known technology has been used but with vastly inflated capabilities, particularly for the time the film was made. Another alternative power source for a stealth ship might be nuclear energy but, in another break with tradition, this Bond villain appears to have no atomic inclinations.

Even with all these adaptations, Carver's ship still won't be invisible, just not as easy to spot by radar or infra-red. If you could get close enough, you would still be able to see it. Even if Carver could borrow the cloaking device Q designed for Bond's car in *Die Another Day*, it would leave a hole in the sea where the hull displaces the water, and a wake behind it when it moves, though the catamaran design would minimise the disturbance of the water. Nevertheless, all these measures cumulatively can be very effective – so effective that a craft can almost be unspottable unless a hatch is opened or a hole blasted in its side with a grenade as Bond does.

With Bond and Wai Lin doing their best to blow up Carver's boat from within and the Royal Navy shelling it from without, it is soon reduced to driftwood, some of which is conveniently big enough for Bond and Wai Lin to clamber onto for the watery kiss that leads into the final credits. Carver is dead, killed by his own sea drill, and left behind as fish food. M dictates a press release to Moneypenny, spinning the truth to state that Carver drowned after an accident on his private yacht, and to suggest that he had died at his own hand.

This is a reference to a third real-life media baron who influenced the Carver character, Robert Maxwell. He was head of the Mirror Corporation that owned a number of newspapers in the UK. Maxwell disappeared from his yacht when he should have been meeting the Bank of England to discuss the £50 million loan he had defaulted on. His naked body was recovered from the Atlantic with only a graze on his shoulder. The poor state of his heart and lungs suggested a heart attack followed by drowning, but there was (unsubstantiated) speculation at the time that he had killed himself knowing that the full extent of his financial misdeeds was about to come to light.[6]

Carver hasn't designed his stealth ship to show off in the traditional Bond villain sense, although he is evidently very proud of his technologically advanced choice of

---

[6] After his death it was discovered he had also defrauded his company's pension funds of millions.

transportation. But other 007 baddies, and Bond himself, move about in ways that can't help but draw attention to themselves.

## Style versus substance

Most Bond villains don't go to the trouble of hiding from Bond, preferring to run away from him, and then only if necessary. Emilio Largo in *Thunderball* goes out of his way to get his luxury yacht, the *Disco Volante* (literally Flying Saucer), noticed. The boat looks as though it is built to impress rather than outpace everyone else but, rather like its owner, it is not all it appears to be. Largo pretends to be searching for shipwrecked treasure as cover for his forays to pick up stolen nuclear warheads, which are loaded and unloaded from his yacht via a secret trapdoor in its hull.[7] When the real plot is revealed, the yacht's sleek outer shell detaches to release a hydrofoil powered by a couple of jet engines. Such features do not come as standard, even on luxury yachts, so the production company had to build one for the film. It cost an eye-watering $200,000, but the boat really did separate in two, even if it took six tries to get it right. Unfortunately for Largo, Bond is already on board the front hydrofoil when he jettisons the back half. At the last moment, Largo is killed and

---

[7] An idea that might have been inspired by Fleming's knowledge of a four and a half thousand tonne Italian tanker, the *Olferra*, moored opposite Gibraltar during the war. Three human torpedoes were captured trying to blow up two aircraft carriers. They claimed they had launched from a submarine but it was later found they had snuck out of the oil tanker via an underwater trapdoor.

Bond jumps overboard before the hydrofoil crashes at high speed into rocks.[8]

Hydrofoils exist and so do luxury yachts; the Bond team simply amalgamated the two into one superboat. Like the cars filled with gadgets, mixing and matching vehicles is yet another way of taking something stylish but fairly ordinary and turning it into something extraordinary. These vehicular chimeras reached their peak in the Roger Moore era. Scaramanga's car with wings bolted on, in *The Man with the Golden Gun*, was not entirely the work of the fevered imagination of the filmmakers. In 1973, two engineers had combined a Ford Pinto with parts from a Cessna to make a flying car, but they came apart again in a test flight, killing both of them. A model was therefore used for the flying scenes in the film. Then there were the totally made-up modes of transport, like the gondola/hovercraft, nicknamed the 'bondola', in *Moonraker*.[9]

The hovercraft/gondola, like much in the Moore films, was played predominantly for laughs. But one of his vehicles stands out as so cool we can forgive its implausibility. Ken Adam's keen eye for good design spotted the Lotus Esprit and thought its sleek narrow frame would look just as good underwater. Nicknamed 'Wet Nellie' in honour of *Little Nellie* from *You Only Live Twice*, audience members are more than willing to suspend disbelief and wish Bond's Lotus car/submarine

---

[8] In spectacular fashion – see chapter 011.

[9] Though the craft was built for real, the skirt didn't inflate evenly, tipping the gondola sideways – and Roger Moore into Venice's Grand Canal five times.

from *The Spy Who Loved Me* could exist. To the credit of the production team, the car/sub was built; in fact, several of them were because they couldn't fit all the gadgets and effects into one car.

One machine was custom-built by Perry Submarines and Oceanographic, using the chassis of a Lotus Esprit (costing around $100,000 at the time, Lotus donated three to the production) and kitted out for underwater work, though it wasn't actually dry inside. The two people 'driving' it had to wear wetsuits and the bubbles you see are from their breathing apparatus.[10] The fins that emerge from the sides were operated by broom handles. The smoke screen was 227l (50gal) of black paint. 'Real' in the world of film isn't quite the same as 'real' in this world and a certain amount of deception is to be expected. For example, the Esprit has a surprising amount of boot space, but how the sea-to-air missile was supposed to fit in alongside the rear propellers is a mystery.

As long as it gets Bond out of a fix, no one is going to worry too much about the ergonomics of the vehicle that does it. In fact, if an escape can be facilitated by an extravagant and/or unusual vehicle it probably will be, even when simpler alternatives would be more expedient.

## Making a spectacle
If you were following someone you suspect to be an assassin, skilled enough to kill two secret service agents, to their home, which happens to be guarded by several

---

[10] They also used a miniature filled with fizzy antacid tablets for some shots.

other heavies more than happy to murder you at any opportunity, it makes sense to have an escape plan. What makes less sense is going to the effort of placing a large and conspicuous jetpack on the roof of the assassin's home and hoping none of the henchmen spot it before Bond needs to use it. But, when you think about it, sneaking a cumbersome jetpack onto a château's roof is not the stupidest part of this plan.

The personal jetpack that makes its appearance in the pre-title sequence of *Thunderball* must be one of the most impractical vehicles in the whole series but, unlike the car-sub, it is a real thing. The Rocket Belt was developed by Bell Textron in the early 1960s for the US Army. It used hydrogen peroxide, which was broken down under high pressure into oxygen and water that emerged as steam from the nozzles. The superheated steam could propel the Rocket Belt's wearer to a maximum height of 18m (59ft), a maximum speed of 55kph (34mph) and a distance of 250m (820ft) during its 20-second flight before the 27kg (60lb) of fuel ran out. The user had to wear specially insulated clothes so as not to get burned by the 750°C (1,380°F) steam. The fact that they are also incredibly difficult and dangerous to fly wouldn't bother Bond, but the pilot who actually flew it in the film refused to take off without a helmet. Hence Sean Connery has to be shown pausing to put one on when he would rather be escaping the very angry henchmen who are after him.

Bond could have run the 250m (820ft) and been halfway back to HQ in the time it took him to suit up, unbuckle and load the pack into his car. Other, improved models have been developed since, but the main problem with all personal jetpacks remains the huge amount of fuel needed to propel something as heavy and un-aerodynamic as a human being any distance. Their fuel efficiency is appalling and they are also as loud as motorbikes. A personal jetpack might be out of the question if you want a stealthy getaway, but it at least has the advantage of being small enough to stow away easily.

If the idea is to travel light, it is difficult to top the one-man helicopter in *You Only Live Twice*. The much-loved *Little Nellie*[11] was designed, named and flown by Wing Commander Ken Wallis. Its precision handling, safe operation and versatility have made it a popular choice for reconnaissance by fictional spies, real-life police and seekers of the Loch Ness Monster. Wallis flew his invention for 47 hours and up to a height of 3,000m (10,000ft) for the film. Though the vehicle really could pack down into a few suitcases, it didn't come with the rockets and guns we see on screen.

As the affectionate names given to some of the vehicles in the franchise show, they have almost acquired personalities of their own. Humans love to anthro-pomorphise everything, and cars, helicopters and gondola/hovercraft hybrids are no exception. These much-loved additions to the franchise have almost become characters in their own right. And, just as Bond is the central figure in the human cast, his car is the star of the mechanical cast.

Deciding on the right car is almost as important, and difficult, as choosing the right actor to play Bond. Both

---

[11] An autogyro rather than a helicopter, because only the rear engine provides power during the flight.

have to be up to the physical rigours of a Bond film and, of course, both have to look good while doing whatever feats are necessary to thwart the latest plot to take over the world. Sometimes fans have been sceptical at the announcement of the new James Bond, only to be won over later when they see how they perform in the role, and the same goes for the cars. The choice of a BMW for Pierce Brosnan's Bond was initially greeted with a certain coolness. But the BMW 750iL in *Tomorrow Never Dies* managed to redeem itself when put into action. Where actors get accessories to help them inhabit the role, so do the cars.

In a brilliant move by the filmmakers, they used a familiar concept – a remote-control car – and made it suitably Bondian. Bond can't see the car he is driving and can only see where it is going due to cameras mounted on the wings sending live images to his phone. The controls for the car are also on the same phone. None of the technology is new, and to prove it the car really was driven by someone who was navigating remotely, though he wasn't using a phone. To keep him out of sight of the cameras, the driver had to lay flat inside the car, unable to see out of the windows and so he was navigating from cameras mounted on the wing mirrors that sent images to screens inside.[12]

BMWs, however, are the exception rather than the rule. Aston Martins have proved to be the most popular

---

[12] It took 17 BMWs to film the chase around the multistorey car park, all with different set-ups, and 15 were wrecked in the process. None of them were roadworthy in any case because of all the modifications that had been made.

choice for Bond to drive, and the DB5 in particular. Introduced in 1964 in *Goldfinger* (see chapter 015), it has made many return appearances for almost no discernible reason other than that it's great. As a private car it is understandable as a fitting choice for a stylish gentleman like James Bond. But why in the twenty-first century he has had Q-branch fit it with a load of 1960s gadgets, and why they have rebuilt it for him at least four times and then allowed him to drive off with it after his retirement from the service, makes no sense. I understand people helping themselves to the odd pen or notepad from their work's stationary cupboard when no one is looking, but would MI6 really let one of their agents disappear with something as big, expensive and dangerous as a vintage car packed with loaded machine guns in its headlights?

Whether it's because of the novelty of a personal jetpack, the ingenious design of a helicopter that can fit in your luggage, or the stylish attributes of the Aston Martin DBS, everyone has a favourite Bond vehicle. The gadgets may be squeezed into unfeasibly small spaces, and the technology slightly more advanced than what is available at the time but, as long as it doesn't strain our sense of believability too far, we are more than happy to go along for the ride.

# 019

# *The World Is Not Enough* and Renard's bullet

To date James Bond has thwarted the plans of 21 self-confessed criminal geniuses. The plans vary but the villains behind them have a lot in common, and not just a hugely inflated opinion of their own abilities. They are ambitious, vain and ruthless. Much of their success in life, cards, backgammon, golf or horse racing is because they cheat. They are usually male and almost invariably not British.[1] They share a similar taste in clothes – Nehru suits and military uniforms are very popular choices. While Bond's relationships with women can hardly be described as healthy, the bad guys rarely share their bed with anyone – and when they do, their partner is usually there under duress.

Several of these megalomaniacs bear the physical scars that can result from living in their violent criminal world. Some, conveniently for Bond, have an unusual physical quirk to help identify them or that can be used against them at some point. For example, Dr No's mechanical hands can't get a grip when the villain

---

[1] Elektra King is the only woman with a grand plan so far in the franchise. King's father is a trademark affable-but-ruthless Brit, but it is emphasised that she takes after her Azerbaijani mother. Scaramanga is British-born but his father was a Cuban ringmaster and his mother a snake charmer of unknown nationality. Elliot Carver is the only Brit who has gone to the bad.

needs to save himself from drowning. Many of these characteristics are shared by the henchpeople in their employ. It's fair to say that Bond villains in every rank of their villainous hierarchy usually fit a stereotype. The physical quirk, the scars and foreign accent have become much-anticipated and parodied characteristics of a Bond villain.[2]

The tropes have become so well known that the franchise has been able to play off them to mislead the audience. Early on in *The World Is Not Enough*, we are introduced to Victor 'Renard' Zokas, a former KGB agent turned freelance terrorist. He has the accent, the scars on his face and a physical quirk – a bullet fired into his head by an MI6 agent that didn't kill him but eliminated his sense of pain and made him stronger. A big neon sign saying 'bad guy' and an arrow pointing at Renard would have been less obvious. But it is misdirection. The real villain, Elektra King, is initially presented as the beautiful victim, but later revealed to be the real brains behind the evil operation.

Renard is no innocent victim – no one with that many clichéd criminal characteristics could be, not in a Bond film. But those characteristics do serve the plot, at least to a certain extent, rather than being used as simple tags to identify the good guys from the bad guys, as Ian Fleming did in the novels.

---

[2] When Mathieu Amalric was cast as Dominic Greene in *Quantum of Solace*, he asked what unusual attribute he was going to get – a dead eye? a terrible scar? perhaps just a terrible haircut? – but was disappointed to be told he would be appearing as himself.

**Defining features**

The villains and henchpeople of Fleming's 007 stories are foreigners,[3] grotesques and cheats. And often all three. He was building on some long-established prejudices and exaggerated them to cartoonish levels. In case the reader is in any doubt that sabotaging missile launches or stealing money are bad things to do, Dr No is given hooks to replace the hands that were cut off in revenge for the money he took from Chinese criminal gangs. Goldfinger must be up to no good because he is short, ugly and his body is out of proportion. Hugo Drax in *Moonraker* is clearly a wrong'un because he cheats at cards. No British gentleman would do that. And, sure enough, Drax is revealed to be a German planning to blow up London with a nuclear missile. There are plenty of worse examples in the books, but you get the idea.

When it came to the films, especially the early ones, a physical defect was little more than cinematic shorthand to identify the bad guy. Bond films move quickly. Characters usually need to be identified as an enemy or ally without lengthy exposition that would slow things down, and several visual shortcuts have been established to do this. Sometimes it is a fluffy white cat cradled in their arms (see chapter 008). Sometimes it is an accessory. In the same way that pirates have eyepatches and bad cowboys have black hats, Bond villains have a hook for a hand or metal teeth. In *Thunderball* it is an eyepatch that shows Emilio Largo is on the side of the pirates rather

---

[3] Perhaps the lack of British baddies is because MI5 is dealing with them.

than the good guys. An almost identical trick is used to pick out the baddie from a crowd of people in *Casino Royale*. On a packed Venetian street a single figure stands out because one lens of his glasses has been blacked out. We don't know what is under the eyepatch or blackened glass, and it doesn't matter to the story, but we know he is a bad guy.

The scars and physical quirks that have been chosen for these Bond villains usually say something about them. They also say something about the extraordinary nature of the human body. It is tough and adaptable. Scars and physical idiosyncrasies are signs of the body's strength and resilience.

Scars are the result of the body's response to damage. A cut to the skin must be quickly covered over to prevent blood loss and block the gap where pathogens could gain access to the body. Initially a blood clot forms, but this is just a temporary structure. Afterwards, collagen is used to bridge the gap and pull the edges of the wound closer together. Collagen is a tough material normally found in the skin to give it its strength and flexibility. In normal skin, the collagen is randomly arranged, but scar tissue forms rapidly in response to the emergency and the body cannot take its time regrowing everything exactly as it was before. The fibres tend to be more closely aligned, meaning scars can feel tight and less elastic than the skin around them. At a later stage the scar tissue will be remodelled by the body, causing it to fade, but after two years there is unlikely to be further change.

The extent of a scar depends on many factors, not just the extent of injury. A wound that takes longer than two weeks to heal usually leaves a visible scar of some kind but they are often little more than a trace of the original injury.[4]

It is noticeable that scars in the world of 007, certainly the permanent ones, are usually given to the bad guys. The most famous of these must be Ernst Stavro Blofeld. The character has been played by many actors both with and without scars, but it is Donald Pleasence's portrayal in *You Only Live Twice*, with a long scar over his eye and down his face, that has become the defining image of a Bond villain.[5] It's the one that was mercilessly parodied in Mike Myer's Austin Powers' films. It has become so iconic that when the character returned to the franchise in *Spectre*, a prequel to the previous films, it is shown how he acquired his signature scar – from Bond blowing up his lair.[6]

Blofeld is not the only scarred character. Alex Trevelyan in *GoldenEye* is left with permanent scars on his face from an explosion that Bond rigged – adding another motive to the long list of reasons why he wants to kill him. Lyutsifer Safin is scarred by the dioxin that was used to poison him. The condition is called 'chloracne', an inflammation from the body launching an immune

---

[4] Some scars can cause itching, pain and anxiety for people who have them, particularly when they are extensive and prominent on the body.
[5] The scar was created by applying glue to Pleasence's face and pinching it into a scar shape when it had dried. He ended up with quite severe bruising down his face.
[6] Losing an eye also means he can get an electronic replacement for when he wants to spy on people, as he does in *No Time to Die*.

response against certain chemicals.[7] As with Trevelyan, it is yet another reason Safin is seeking revenge on the people who have hurt him.

The six cinematic Bonds pick up a few scratches on their adventures but only Daniel Craig's Bond retains a scar from one film to the next.[8] This is in contrast to the novels, where Bond is said to sport a 3in (equivalent to 8cm) vertical scar on his right cheek.[9] Bond's invulnerability is renowned but ridiculous (see chapter 021). The main function of a scar in a 007 film remains a means of identifying the villain. And, although all scars are different, Bond sometimes needs more help to identify reclusive, camera-shy criminals and assassins.

## Distinguishing marks

Some of the physical features given to Bond villains are a little surprising. A few seem to have been dreamt up merely because villains are supposed to have some distinguishing feature. Stromberg's webbed fingers in *The Spy Who Loved Me* link him to his preferred underwater lifestyle but have no benefit to the plot and are scarcely even referenced during the film.

A little extra skin between the fingers or toes is known as syndactyly and is not that unusual. As the foetus grows in the womb the hands and feet are webbed. At around six weeks' gestation, the cells that

---

[7] The film is perhaps deliberately referencing the real-life poisoning of Ukrainian opposition leader Viktor Yushchenko in 2004.

[8] The scar from where he was shot by Patrice and Moneypenny in *Skyfall* is still visible in *No Time to Die*, two films and 11 years later.

[9] He also has a skin graft on the back of his hand to cover a distinguishing scar left by a SMERSH agent.

form the webbing are destroyed to leave unconnected toes and fingers. In some people, the programmed cell death doesn't always go to plan and traces of webbing are left behind.[10]

In a similar vein, Scaramanga's third nipple in *The Man with the Golden Gun* is not an uncommon feature. The fact that men have nipples at all can seem odd but is again down to the human foetus developing in a certain way. All of us start off with female characteristics, which, in boys, are later overridden by a flood of male hormones. If the body contains the genetic instruction to create nipples in the first place, these instructions can be repeated. Extra nipples, or pseudomamma, usually follow the 'milk line', starting at the armpit and ending at the groin.[11] Though they are sometimes thought to be a sign of fertility, they are actually an atavism or evolutionary hangover from our ancestors. Scaramanga's bonus feature serves no purpose other than to identify him. Little is known about the rest of his appearance, and so Bond can impersonate Scaramanga to gain some information simply by sticking a prosthetic nipple to his chest.[12]

Le Chiffre's identifying feature in the film version of *Casino Royale* also proves useful to Bond. His eye weeps blood when he is stressed, so the audience and Bond have a useful visual clue to his state of mind. Bond can also use it as a 'tell' when he and Le Chiffre are playing

---

[10] More complicated forms of syndactyly can involve merged finger joints, which are more difficult to separate surgically, but Stromberg appears to have the simple kind involving just skin.

[11] Pseudomamma can appear as far away as the foot.

[12] In the wrong place.

against each other in a high-stakes poker match.[13] Haemolacria is a genuine medical condition that can mean tears are stained with blood. It can be caused by infections, such as conjunctivitis, hormone changes or physical injury. Le Chiffre appears to have scarring around his eye and damage to the eye itself, suggesting injury. With such an obvious tell, it's surprising Le Chiffre hasn't had his haemolacria corrected, which is certainly possible with surgery. Other Bond villains have been more than willing to go under the knife if it furthers their evil plans.

## Means of disguise

In *Diamonds Are Forever*, Blofeld subjects a selection of his minions to extensive plastic surgery so they will look like him and confuse Bond as to which of the several criminal masterminds he is presented with should be killed. In *On Her Majesty's Secret Service*, Blofeld goes under the knife himself to have his earlobes trimmed. In an effort to gain the superior status he feels is his due, he wants to prove he is a member of an aristocratic family, all of whom have missing earlobes because of a genetic quirk.

One Bond baddie, however, went further than all the others. Colonel Tan-Sun Moon transforms himself into Gustav Graves, not by plastic surgery but by genetic manipulation. Gene therapy was the hot topic of the day when *Die Another Day* was produced in 2002. The

---

[13] In the novel, Le Chiffre had no such injury but was overweight and regularly using a Benzedrine inhaler. In the film, he lost the weight but kept the inhaler.

Human Genome Project was 12 years into its ambitious task of creating a complete map of the human genetic code.[14] The film managed to take several aspects of genuine genetic science and contort them into one outlandish plot.

Moon's treatment involves destroying his bone marrow to wipe his genetic slate clean, and then starting again with new DNA harvested from healthy donors. Destroying the body's bone marrow and replacing it with a donor bone marrow is a medical procedure that has been carried out since the 1950s. It has been used to treat leukaemia and other blood disorders, but only affects cells that are made within the bone, which is admittedly a lot, but by no means all of them and none that would change a person's outward appearance.

Almost every cell in the body carries a complete set of DNA instructions.[15] New cells are generated by one cell copying its DNA contents and using those genetic instructions to make an identical cell that splits off from the original. In other words liver cells beget more liver cells and kidney cells beget more kidney cells etc. To achieve the complete physical transformations shown in Die Another Day, the patient would need to replace the DNA in more than just their bone marrow cells. Then there is the problem that the body checks for foreign matter, and anything it doesn't recognise as 'self' is

---

[14] A rough draft was published in 2000 and the project reached completion a year after the film was released.

[15] Red blood cells have no DNA and sperm and egg cells each contain only half a set.

destroyed. It's why bone marrow and organ donors have to be a close genetic match or the recipient's body will reject it. Moon's genetic refurbishment wouldn't just be painful, as the film states, it would probably kill him.

Rather than going for wholesale replacement, what if Moon's existing code could be edited in some way? Since the 1980s vectors like viruses have been used to infect cells and insert new short sequences of DNA with varying degrees of success. A whole new world of gene editing opened up in 2012 with the development of CRISPR techniques.[16]

CRISPR (clustered regular interspaced short palindromic repeats) are DNA fragments taken from organisms that don't have a cell nucleus, such as bacteria. The bacteria use the fragments to delete foreign material from their DNA as part of their defence system. Combining the fragments with enzymes that recognise specific sequences of DNA can be used to snip out or insert specific genes in individual cells. It has huge potential, from boosting crop yields to modifying mosquitoes to prevent the spread of malaria, but it can't completely transform an adult's physical appearance.

The problem with any gene editing process is the complexity of DNA. A single sequence of DNA doesn't always code for a single feature, and much of the genetic code within our DNA controls how much or how little of any particular sequence is expressed. It is no good inserting DNA that never gets used.

---

[16] By Jennifer Doudna and Emmanuelle Charpentier, who were awarded a Nobel Prize in Chemistry in 2020 for one of the most significant discoveries in the history of science.

If the goal is to change your appearance then plastic surgery is the only real option. Painful, genetic tinkering that robs you of sleep sounds awful and won't work anyway.[17] Many people are interested in using CRISPR techniques to improve certain traits, perhaps to make them stronger or more resilient to infection. The technology is still in its infancy, but enhanced strength or resilience would certainly be more realistic as well as in keeping with Moon's fellow villains.

## Playing to your strengths

The first henchman to be given the now characteristic attributes of enormous strength and imperviousness to pain was Red Grant in *From Russia with Love*. He doesn't even flinch when Rosa Klebb punches him in the stomach with a knuckle duster. In the inevitable fight between the main heavy and Bond, we know 007 will have to use his brains to overcome his opponent's brawn. In Red Grant's case, a Q-branch-adapted briefcase releases knockout gas into his face, incapacitating him.

Many actors have been chosen to fill the role of henchman for their physical size. Harold Sakata, a professional wrestler, was cast as Oddjob in *Goldfinger* based on 'the sheer tonnage of the man'. It is a tradition that peaked with *The Spy Who Loved Me*, and one of the most beloved baddies in the whole Bond series, Jaws. Portrayed by the 7ft 2in Richard Kiel, his impressive height was due to a medical condition, acromegaly. A

---

[17] I'm sure the idea of such a procedure would keep you awake at night, but why it stops the patient from physically being able to sleep is a mystery.

benign tumour growing in the pituitary gland causes
growth hormone to be released long after most adults
have normally stopped growing. It can result in large
hands, feet and facial features. Keil's already imposing
presence is made even more menacing by the addition
of metal teeth.[18] Over two films,[19] Jaws is seen biting his
way through padlocks, steel cables and people's necks.

Although metal teeth would be strong and could be
sharpened to give a better cutting edge, they wouldn't
increase Jaws's bite strength, as that comes from the
muscles attached to the jaw. The most powerful human
bite was recorded in 1986 by Richard Hofmann at a
strength of 975 psi, six times greater than the average. To
put that in the context of Jaws, a cable car cable can
withstand a grip of 30,000 psi without being damaged.
In the film, the cable, and anything else Kiel had to bite
through, was made from liquorice. Even so, the teeth cut
into Kiel's gums so he couldn't actually bite down with
any pressure.[20] The same metal teeth also prove to be
Jaws's downfall. They get stuck to magnets and are
excellent conductors of electricity, allowing Bond to
escape his enormous clutches.

Another overly strong, but averagely sized, baddie is
Max Zorin in *A View to Kill*. He was born as the product
of experiments carried out by Hans Glaub (who changes

---

[18] The teeth were inspired by the steel-capped teeth sported by one
of the henchmen, Sol 'Horror' Horrowitz, in Fleming's novel *The
Spy Who Loved Me*.

[19] Jaws was so popular with fans he returned in the following film,
*Moonraker*.

[20] The prosthetic teeth also made the actor feel ill and Kiel could
only wear them for short spells.

his name to Carl Mortner to evade justice) on pregnant women in the Nazi concentration camps.[21] It was hoped the steroids they were given would result in babies with enhanced intelligence. Zorin grows up to be very intelligent and very strong, but also psychopathic. Hence he is more than a physical match for Bond and has no compunction about killing him, or thousands of other people, to get what he wants. In fact, he rather enjoys it.

Steroids are biologically active molecules with a certain basic chemical structure. Some are hormones like testosterone and are well known, but there are many others involved in various important processes that keep us healthy. Different modifications to the steroid structure mean these molecules interact with different receptors in the body, causing different effects. The basic steroid structure can also be modified artificially to create new steroids that can affect the body too.

Steroids were first made in a lab by the German biochemist Adolf Butenandt in the 1930s. During the Second World War he conducted medical research into helping fighter pilots function better at high altitudes. But stories of steroids being given to German soldiers to increase their strength and bravery have never been proved and seem unlikely. As discussed in *A View to a Kill*, medical experiments were carried out in the Nazi concentration camps. Given the horrific nature of those

---

[21] The character may have been based on Karl Clauberg, a gynaecologist who worked alongside the notorious Josef Mengele in the Auschwitz death camp, though his focus was on sterilisation rather than the possibility of breeding a race of super fighters.

human experiments, it is not much of a stretch to believe attempts were made to create a super-race, although there is no evidence they were. Even if such experiments had been carried out, there is no reason to think they would have resulted in anything like Zorin. His backstory is simply there to provide some sort of explanation for his super strength and evil actions. It's nonsense, but it is more than most Bond villains get.

Though Bond has no chance of defeating these extra strong baddies one-on-one, they all meet their maker one way or another. Bond and Zorin's final showdown is conducted on top of the Golden Gate Bridge in San Francisco. Bond doesn't have to deal with Zorin's immense strength, as the villain slips and falls to his death. Oddjob (*Goldfinger*) is electrocuted; Stamper (*Tomorrow Never Dies*) is obliterated by the exhaust end of a missile; Mr Kil (*Die Another Day*) gets a laser through the skull, and so on.

Where most of these physical or mechanical additions are supposed to give the bad guys an unfair advantage, Renard's bullet, in *The World Is Not Enough*, would seem to be a disadvantage. The missile is working its way through his brain matter and will eventually kill him. All Bond need do is wait. Except that, knowing his time is limited, Renard is set to do as much damage as possible before the bullet claims his life.

**Strengths or weaknesses**
The one advantage the bullet appears to have given Renard is that as it slowly moves through his brain, it destroys his ability to feel pain. Pain is a complicated thing. There are specialised nerves and nerve endings in

almost every body tissue that are stimulated by irritating chemicals, pressure and muscle contractions. Signals from these nerves are transmitted to regions of the spinal cord. The activity of these nerves can be measured but the effect they have on a person – how an individual perceives pain and responds to it – is entirely subjective and varies enormously. What is known is that damage to specific nerves in the spinal cord can permanently impair pain sensation. But Renard's bullet isn't in his spine.

Pain signals from the spinal cord are passed on to the brain via the thalamus and brainstem. Damage to these important pain pathway hubs would potentially result in more than just a diminished response to pain. There could be dramatic changes in personality from damage to the thalamus. As the brainstem controls all basic bodily functions, such as breathing and heartbeat, knocking out this region has some potentially very serious consequences. But there are also lots of other parts of the brain involved in pain perception.

The brain is very interconnected, though some regions are tasked with specific roles. The regions of the brain that are activated by pain have been investigated using a technique called functional MRI, or fMRI. While a person carries out a specific task, an MRI machine can monitor which parts of the brain have increased blood flow, suggesting which areas are more activated by the task.[22] And it turns out there are a lot of regions in the brain that are activated by pain.

---

[22] Damage to the brain through strokes or physical injury can also reveal which areas are responsible for what by observing the effects on the patient.

In the 1950s patients were given frontal lobotomies to treat severe pain. They felt pain just as before the operation but their anguish and suffering was eliminated. This might be considered a success, but the surgery had other unintended consequences, including severe personality changes. The complications and impairments suffered by the patients were too severe and the procedure has since been abandoned. If surgery can have this much of an effect, imagine what a bullet can do.

Bullets are not precision tools suitable for brain surgery. They do not bore nice, neat tunnels through the brain, as shown in the film. If the projectile has enough energy to smash through the skull, soft brain matter poses no obstacles (see chapter 009). Damage will be extensive and likely to knock out several other important functions along with any regions involved in pain perception.

The brain presumably goes to so much trouble to distribute these areas because pain is important. It is a way of alerting the body to danger. Without it you would be completely unaware of serious damage to the body. Renard may appear to be getting stronger, but losing his sense of pain also means he loses the aches and twinges that are early warning signs his body is under strain. His high-risk lifestyle as an international criminal, getting into fights and handling radioactive substances, would mean he runs the risk of shortening his life still further. He finally dies from a plutonium rod through the chest. It is a horrific injury, but if the bullet has done what the film claims, he would feel nothing.

Renard's physical quirk, like all the others, is a huge exaggeration for the benefit of the fantasy world of 007.

Bond is only as good as the villains he is pitted against. Overly ambitious and convoluted plans for world domination aren't always enough. The contrast between the good-looking, British, romantically successful secret agent and his enemies doesn't just need to be stark, it is highlighted, underlined and surrounded by easily understood symbolic references. Subtlety is not the franchise's strong point or the reason audiences love it.

# 020

# *Die Another Day* and being sucked out of a plane

It's not easy being a Bond villain. They have a lot to worry about – driving challenging building projects, developing large-scale and complex new technologies, high staff turnover, British secret service agents turning up unexpectedly to ruin their day. The over-complicated lifestyle they have chosen means much of their stress is self-inflicted. Sometimes the pressure, or lack of it, is too much.

In a series of films known for finding inventive ways to kill their characters, a few fatal themes have emerged over the years: the misapplication of electrical power (see chapter 012); animal intervention (chapter 008); and through pressure – either too much or too little. Dr Kananga, Hugo Drax, Milton Krest and Admiral Chuck Farrell have all been squeezed or stretched to death. Gustav Graves and Auric Goldfinger have both been pressured into taking paths they would rather not, out of a plane and into the thin air outside via a window. While on the face of it these deaths seem ludicrously implausible, they are scientifically not so far off the mark. Fortunately, very few of us are ever likely to find ourselves in their uncomfortable situations.

**Frequent flyer**
Flying is still the safest method of transportation, but James Bond seems to have had a particularly bad run of

luck over the past 60 years. Though he spends a lot of time in aeroplanes, he rarely gets to sit back and enjoy the in-flight movie or sample the airline food. Given how often people get sucked or pushed out of planes in 007 films, it would make me think twice about boarding an aircraft if I spotted Bond at the check-in desk. His supreme talent for destroying almost anything he touches is particularly worrying when travelling in a vehicle where structural integrity has a slightly bigger significance than land-based transportation.

Most passenger aircraft fly between 9,000 and 13,000m (30,000 and 42,000ft), where the air is considerably thinner than at ground level. If a plane is flying at 9,000m, the outside pressure is 0.3 atmospheres, much lower than the one atmosphere pressure at sea level. Therefore, the inside of these planes is pressurised, at around 0.6 atmospheres,[1] to allow the passengers and crew to breathe normally. The plane is not hermetically sealed because this would also cause problems over time as the oxygen inside the craft would soon be used up. Instead, there are valves at the front and back, and as the plane moves forward, air is drawn in the front valve and out through the rear. The size of these inlet and outlet valves are constantly adjusted to keep the air inside the plane fresh and at a constant pressure.

Creating more holes in the external structure of an aeroplane means the higher-pressure air inside will rush out. This sudden depressurisation has been used to suck several villains out of their plane and to their deaths. In reality, in the unlikely event of a sudden leak on a plane,

---

[1] Roughly equivalent to being 2,130m (7,000ft) above sea level.

the inlets and outlets can adjust to keep the pressure at a healthy level. Real problems only occur if the leak is too big to compensate for, but how big does that leak have to be?

In *Goldfinger*, Bond is on board Goldfinger's private jet, a Lockheed Jetstar, and Pussy Galore is pointing a gun towards him. He says that firing the bullet would create a hole in the fuselage, the plane would depressurise and they would both be sucked out into space. He's exaggerating (see chapter 009), but it stops Galore killing him. It also explains to the audience what to expect if a hole is suddenly created in a plane, as it is later when Goldfinger sneaks aboard the flight carrying Bond to meet the President of the United States.[2] He brandishes his gold-plated gun at Bond and in the ensuing struggle it goes off. If the bullet had gone through the fuselage it would have made a small hole and not much difference to the pressure inside the cabin, but it hits a window and the glass smashes, creating a big hole. Anything not bolted down is sucked outside. Bond holds on to the furniture for dear life, but Goldfinger cannot get a grip and he is squeezed out the window like toothpaste out of a tube.

In the unlikely event that a window on a commercial aeroplane failed mid-flight, there would indeed be a rapid decompression, the air would be sucked out of the cabin, and the lungs of the passengers on board. A mist would form as liquids, bodily fluids and the contents of the drinks trolley rapidly vaporised. It would get very cold very quickly. Anyone sitting next to the missing

---

[2] This time he is on board a Lockheed US VC-140B.

window would be sucked out of their seat – even a seat belt wouldn't hold them. However, the windows at the sides of a plane are smaller than the average shoulder width of an adult man, and the passenger – unless they are very small – would most likely be jammed against the window, effectively blocking the hole. Goldfinger is a well-padded bad guy and should not have been able to squeeze through the window. On a normal passenger plane, the fog would clear, oxygen masks would drop from above and the plane could be landed safely.

The oxygen masks are important. Mild to moderate altitude sickness starts between 2,400 and 3,600m (8,000 and 12,000ft) because there is less oxygen than at ground level. Nausea, dizziness, fatigue, confusion and shortness of breath do not help when you are trying to fly a plane. And things get worse the higher you go. Galore and Bond, who are still on the plane after Goldfinger has made his exit, don't bother with oxygen masks but they are rather preoccupied. The hole in its side has apparently caused the plane to go into a steep dive and Galore and Bond's combined efforts can't pull them out of it. They parachute to safety and leave the plane to crash in a ball of flames.

Incredibly, a not dissimilar event happened on a real aircraft in 1990, but with a much less dramatic outcome. The British Airways flight was climbing past 6,000m (20,000ft) when one of the cockpit windows fell out. These windows are much bigger than the ones passengers get to look through. Every loose item in the cockpit was sucked out of the hole and even an emergency oxygen bottle that had been bolted down flew past the co-pilot's head, nearly knocking him out. The pilot, Captain Tim

Lancaster, was sucked out of his seat belt and through the window. A flight attendant, Nigel Ogden, managed to grab hold of him but started to be sucked out himself. Another member of the crew grabbed hold of Ogden's trouser belt and tied it to the pilot's shoulder strap to stop him slipping further. Meanwhile, the pilot's body became bent double through the window, stopping him being sucked any further and partly blocking the hole. His arms were flailing and his face was banging against the windscreen, with blood pouring from his nose. The co-pilot managed to land the craft 18 minutes later. Captain Lancaster suffered nothing more serious than frostbite and a few broken ribs.

Goldfinger's demise and the subsequent plane crash are far from credible, but a few decades later the 007 team tried again. The 2002 film *Die Another Day* marked both the 40th anniversary of the Bond franchise and its 20th instalment. The filmmakers therefore crammed as many in-jokes and references to previous films as they could into the 2 hour 13 minute runtime. Gustav Graves' death is an homage to Goldfinger's, but presented in a slightly more plausible way, which is surprising given the implausibility of the rest of the film.

This time we have the villain flying on his custom-designed An-124-100. The plane is big enough to have a well-equipped gym, as well as carry a few very fancy cars and a helicopter inside. Bond is also on board for the final battle to stop Graves and save the world. The plane has some very large windows, easily big enough for a person to fit through, as demonstrated by Graves' tech-support man, Vlad, who is the first person to get sucked out when one of the windows is broken. Two more henchmen follow Vlad outside, but Bond and Graves manage to hold on until the pressure reaches an equilibrium. The fight between Graves and Bond continues, as does the disintegration of the plane, and more windows and panels are lost. Bond finally gets rid of him by pulling the cord on Grave's parachute, which opens due to the considerable breeze generated inside the plane thanks to the many holes it now has. The parachute pulls Graves out and into a jet engine. Bigger holes cause bigger problems.

In 1988, a jetliner was making its way to Honolulu. At 7,300m (24,000ft), a section of the first-class cabin was ripped off, leaving a 5.5m (18ft) hole. A flight attendant was swept off her feet and out of the aircraft. Of the other passengers and crew, 61 were treated for minor injuries, but the plane landed safely with no further casualties. Less than a year later a much larger hole (3 x 12m/10 x 40ft) appeared in the side of another plane flying from Los Angeles to New Zealand. This time, 11 people died and 16 others were treated for mostly minor injuries. The remaining 327 on board survived and the plane landed safely in Hawaii, even though only two of its four engines were still working. It is surprising how much damage an aeroplane can sustain and continue to fly. The rapidly disintegrating An-124-100 in *Die Another Day* continues to fly long enough for Bond and his CIA accomplice, Jinx, to escape in the helicopter stowed in the plane's hold.

Suddenly exiting a plane at high altitude is far from ideal but, if you are strapped to a parachute or can clamber into a helicopter before you leave, your chances of making it to the ground safely are greatly improved (see chapter 010). However, your chances of survival diminish with distance.

## Under pressure

There is no line or distance that marks the border between the Earth's atmosphere and space. But the further you move away from the Earth, the thinner the atmosphere gets. By the time you are 100km (62 miles) above the surface of the Earth, the air is too thin to support an aircraft unless it is flying at orbital speed. So this is as good a place as any to describe the boundary between the atmosphere and space.

The 1979 Bond film *Moonraker* took 007 well beyond the 100km mark and into outer space in search of the villain Hugo Drax. Drax has built an enormous secret space station that is obviously a lot further away from the Earth than 100km (see chapter 005). The closest real-life approximation to his secret space lair would be the International Space Station (ISS), which orbits 400km (249 miles) above the Earth. At this distance there isn't any atmosphere, or much of anything at all, to speak of. Bond catches up with Drax inside the spacecraft and uses a dart fired from his wrist to force Drax into an airlock. The airlock is then opened and Drax is sucked out into the vacuum of space.

When real-life astronauts venture outside their pressurised space vehicles they wear carefully designed, sophisticated spacesuits, and with good reason. Drax is wearing nothing more than standard villain attire, a Nehru suit, and is subjected to the full consequences of being surrounded by absolutely nothing. The first thing he would notice would be a chill as all the sweat from his body instantly evaporated. All the liquids on the outside of his body, including tears and saliva would vaporise within seconds, causing a stinging sensation. There would also be the sudden rush of air out of his lungs. He would likely lose consciousness around 10 seconds into his ordeal, giving him only a very short window of opportunity to try and get back inside. If he could have done this, he would have roughly two minutes to repressurise and recover. We know this because of an accident in a NASA testing lab.

In 1966 a NASA technician was in a vacuum chamber testing a spacesuit. Unfortunately, no one had noticed that the suit had a faulty hose attachment. Though he didn't have to suffer the extreme cold of space, which is close to absolute zero, he was unprotected in a vacuum for 87 seconds before the chamber could be repressurised. Ten seconds after the start of the experiment he fell unconscious but, apart from earache from the rapid pressure changes, he was otherwise unharmed. Drax was not so lucky.

Being unconscious after just 10 seconds is probably for the best, considering what would happen next. Drax's body, and everything within it, would initially be at normal atmospheric pressure, but not for long. Any pockets of gas within the body would escape via the nearest exit. And any that can't find a way out would expand. There is also a lot of water inside a human body; it makes up between two-thirds and three-quarters of our mass. As pressure drops, so does the boiling point of liquids. Even in the cold of space there will be enough heat in the body itself to cause the blood to boil. Drax's skin would stretch and the body swell up like a balloon until it ripped apart. Instead of grisly realism that would have got the film an adult certificate, the filmmakers chose to show Drax drifting away, intact and uninflated, with just a surprised look on his face. When they chose to show a similar scenario in a later film, the pendulum was swung perhaps a little too far the other way.

In *Licence to Kill*, Milton Krest is the owner of an aquarium that acts as a front for the main villain, Sanchez, and his drug smuggling. Krest has a boat he uses to do research and catch the exotic specimens he sells. This boat is equipped with a hyperbaric chamber for divers who return too quickly from working at great depths.

Dives must be carefully managed to prevent gases dissolved in the blood from expanding as a diver approaches the surface. These bubbles can cause several problems, depending on where they form, though they can also move through the body. Most common is into the joints, causing pain. A more serious situation occurs if the bubbles form or move to the arteries, brain or spinal cord. Treatment is given in a hyperbaric chamber. These chambers are big enough to enclose a whole person, and oxygen-enriched air is forced inside at higher than normal pressure to mimic the effects of diving. The increased pressure helps to redissolve the bubbles and the increased oxygen helps recovery in those parts of the body that might have had the blood supply cut off by a bubble blocking the way. During the process the person continues to breathe normally, expelling the excess gas from their lungs as the pressure is slowly lowered to prevent the bubbles reforming.

In the film, the main villain, Sanchez, convinced that Krest has double-crossed him, pushes him into the hyperbaric chamber. Sanchez turns the dial, increasing the pressure to dangerously high levels. As Krest holds his ears and begs for mercy, Sanchez cuts the oxygen feed with an axe. The air rushes out of the chamber, causing Krest to explosively decompress.

The special effects team spent ages making a rubber head that looked like the actor Anthony Zerbe and filled it with fake blood. Unfortunately, all their efforts alarmed the censors, who forced them to cut the sequence to the bare minimum.[3] All we see in the film is the head swelling up like a balloon through the chamber's porthole and then the glass suddenly goes red. It's dramatic but unrealistic. There are plenty of small blood vessels and weak points in Krest's body that would have given way before the whole head, but that would probably have been even more distressing to see. When a similar special effect was used in *Live and Let Die*, it was blood-free and even more unrealistic.

The 1973 film sees Bond captured in Dr Kananga's lair and, over drinks, Kananga asks about a weapon they recovered – a rifle with unusual bullets. Bond explains it is a shark gun and the bullets contain compressed gas. Why you would use compressed gas to kill a shark, I don't know. If you are hunting sharks, surely you would want to have a trophy to show for your efforts, not unidentifiable scattered shark meat. If you are using it to defend yourself against a shark attack, shredded shark in the water is only likely to attract more sharks. Nevertheless, Kananga is impressed

---

[3] And it was still considered graphic enough to earn the film a much higher certification than most of the others in the franchise.

by the weapon and fires one of the bullets into a leather chair that inflates and explodes, much to the dismay of Whisper who was sitting on it at the time. The audience has been primed, so no further explanation is needed when Bond and Kananga are later struggling in a shark-infested pool and Bond forces Kananga to swallow one of the bullets.[4] The compressed gas is released and a rapidly expanding villain floats up to the ceiling before exploding. Rather than a shower of pink mist, there is nothing left of him other than a few shreds of his clothes.

Instead of inflating from the outside, as Krest did, Kananga is forced to expand from within. The sudden release of compressed gas inside a human body would undoubtedly cause appalling injuries, but the gas would initially be concentrated in the stomach area. What we see is all of Kananga inflating, something that should only happen if the gas had been absorbed throughout the body. Floating up to the ceiling is also improbable as the gas is just ordinary air, the same weight as the air everywhere else. Kananga should not have been so buoyant, but I'm picking holes in one of the more memorable deaths in the franchise.

Another memorable death is that of Admiral Chuck Farrell in *GoldenEye* because, as far as deaths go in the franchise, his might be the most enjoyable. Where plenty of characters have been stretched beyond their physical limit, Onatopp's victims have the opposite problem.

---

[4] How the bullet is fired from inside Kananga's stomach, I can't explain. Any ammunition that unstable would be unusable.

## The big squeeze

Onatopp likes to get up close and personal with her victims. At the most intimate moment, she wraps her legs around their chest and squeezes them until they die. This is actually quite difficult to do. She must have had a lot of practice.

The chest needs to expand a few centimetres for air to be drawn into the lungs, so it is certainly possible to kill someone by preventing that from happening. Accidents on building sites have left people buried chest-deep in sand, soil or gravel and unable to breathe. There are only a few short minutes available to try and dig someone out before lack of oxygen kills them.[5] What is surprising is just how much weight an adult's chest can withstand.

In 1692, in Salem, Massachusetts, Giles Corey was accused of witchcraft but, when asked if he was guilty or not guilty, he refused to answer. The punishment for his silence was to be pressed to death. In the sixteenth and seventeenth centuries this was the standard punishment for refusing to enter a plea. The individual would be taken to a special room or yard within the prison, where they would be laid on the floor and have weights piled on top of them until they died. As an act of mercy a sharp stone might be placed under their back so the weight would break the spine and hasten death. Giles Corey was not so fortunate. There was apparently no stone under his spine when 180kg (400lb) of stones were loaded onto a board placed on his chest. It took

---

[5] Another serious problem in this situation is compressing veins so that blood can't circulate properly around the body.

two days for him to die. His last words were allegedly, 'more weight'.

If Onatopp is using her thighs to crush her victims, either she can exert more than 180kg of pressure on the chest, or something else is going on. Maybe she is able to angle a heel or toe into the spine to break it, as the stone did in the past. Or perhaps because she is not applying the pressure evenly, a few ribs are being cracked in the crushing process. Broken ribs would make it more difficult for the chest to expand, and any sharp pieces of bone might puncture the lungs. However she does it, it is an impressive, if macabre, talent.

Being crushed to death by a woman would be a strangely appropriate end for James Bond. But when Onatopp tries to put the squeeze on him he is able to resist. Maybe his pride and self-assurance helps swell his chest while he fights her off. As with most of his sexual relationships, he is able to untangle himself and move on, apparently unaffected by the experience.

# 021

# *Casino Royale* and the knotted rope

James Bond has a remarkable body. Though not able to leap tall buildings with a single bound or run faster than a locomotive, he has an ability to defy the odds that only fictional characters can achieve. Sixty years of being drugged, poisoned, beaten, chased, dropped, crashed, burned and tortured has proved his resilience to physical and chemical damage beyond any real-life human ability. It seems just as difficult to dent his mental health, but that is a discussion for chapter 024.

We never see Bond attending what must be necessary and regular check-ups at sexual health clinics, or wryly commenting on hospital food during lengthy recoveries from his injuries. Given that Bond is captured on almost every mission, sometimes embarrassingly easily, his resistance to interrogation and torture is essential. He has been put through some extraordinary scenarios to try and persuade him to do what the villain wants – be it to reveal information, come over to the dark side or just die horribly. Sexually harassed by Silva (*Skyfall*), scraped over coral reefs by Kristatos (*For Your Eyes Only*), brain surgically destroyed by Blofeld (*Spectre*)[1] – none of it

---

[1] A scene taken from *Colonel Sun*, written by Kingsley Amis. It is the only non-Fleming Bond novel to be used as source material for the films.

seems to have done any permanent damage to Bond, or even made much of an impression on our tough hero.

Without his incredible feats of endurance, it's a wonder MI6 would trust him any further than the local coffee shop with anything more secret than how many sugars M takes in her tea. No matter what, Connery, Moore, Dalton and Brosnan brush the dust from their shoulders, straighten their tie and emerge fresh and ready for their next adventure. To help him the filmmakers gave Bond a convenient short-term memory problem. Each adventure operates within its own bubble, with some of the same recurring characters, but no recollection of previous painful experiences. Things changed in the Daniel Craig era; the characters retained memories of events from one film to the next. One scene in particular certainly stuck in the memories of those who saw it – Bond tied to a chair and tortured with a knotted rope.

The 2006 film was something of a reboot for the franchise, showing James Bond acquiring his double-0 status and setting out on his first mission. In *Skyfall*, only two films later, questions are being asked about his fitness for the field. But is this any surprise given what he must go through as part of his day job? *Casino Royale* has Bond chase down a bomber on foot through, around and over a construction site, being shot at, nearly blown up, beaten, poisoned, involved in a car accident, brutally tortured, shot at and beaten again and nearly drowned in a collapsing building. It's a lot. Anything is possible in fiction, but how far did Fleming and the film producers have to stretch credibility to ensure Bond survives to spy another day?

## Preparation

Any specialist job requires training. At the most basic level a double-0 agent needs to be physically fit, an excellent marksperson[2] (see chapter 009), be able to drive, fly or steer a huge range of vehicles (chapter 018), and be well versed in explosives of all kinds (chapter 011) as well as gadgets (chapter 002). While skill is undoubtedly involved, a lot can be learned and practised to reach a sufficient competency, and the competency for a double-0 agent would be expected to be very high. Though Bond has a background in the navy, he would need to go through every form of training in all branches of the military to stand a chance against the diverse dangers he faces on each mission. How else would Bond survive being spun round in a circle without going through pilot training in the air force?

The centrifuge trainer in *Moonraker* has been built by Hugo Drax so his astronauts will be able to tolerate the extreme forces experienced during shuttle launches. These machines are also used in real life for astronauts and to train pilots who may experience high g-forces (g) when flying very fast and making sudden changes of direction. The machine spins round in a circle, accelerating the person inside it so they feel a force pushing them backwards. As the acceleration, or force, increases, soft tissue deforms and blood can be pushed to the lower parts of the body, away from the brain, depriving it of the oxygen the blood is carrying. Most people black out experiencing between

---

[2] According to the films, the double-0 section has had women in its ranks since at least 1965 and the Thunderball operation.

4 and 6g (four to six times the force of gravity). A constant application of 15g for more than a minute can be deadly but a person can increase their tolerance with training. A g-suit, with tight-fitting trousers and built-in bladders that can be inflated to increase the pressure on the legs and abdomen, also helps prevent blood from pooling, and can also give an extra g of tolerance to the wearer.

Drax's human centrifuge can generate up to 20g on any unfortunate occupant.[3] Bond is strapped into the seat, wearing a suit that is certainly made to measure but definitely not a g-suit, and the machine starts to speed up. At 7g, Bond realises the emergency stop switch has been disabled. At 13g, when most people would be unconscious, he is still awake and thinking clearly enough to stop the centrifuge by firing one of his wrist darts into the controls.

Covering every eventuality in a double-0 training course would be impractical, especially given the fiendish ingenuity of the villains they are likely to be up against. But perhaps a few bonus lessons on disarming nuclear devices, shark defence and piranha avoidance wouldn't go amiss. There are, however, some things that are difficult if not impossible to train for, such as unexpected or life-threatening situations.

The military uses training exercises, putting its combatants through near-real combat situations, to prepare them for what they might encounter in the field. It is much easier to deal with a threat having been

---

[3] Presumably built to such a high, unsafe standard so it can double as a means of killing nosy secret agents or uncooperative henchmen.

exposed to it, or something like it, previously. We see
Bond taking part in these kinds of training exercises in
*The Living Daylights* and *Die Another Day*.

Even interrogation techniques can be rehearsed, but
there are, of course, limits on what can be done ethically
for training purposes. Part of SAS (Special Air Service)
preparations involve putting its members through hours
of isolation, stress positions and blaring white noise, as
something approaching what they might expect to
experience should they be captured. Showing this in a
007 film would not make for good cinema, though
other aspects of their training – such as the intense
questioning, with a few Bondian flourishes like
scorpions, lasers or mind-altering chemicals[4] – have
made it on screen. And the prospect of making paying
customers squirm in their seats hasn't stopped the
filmmakers from showing extensive torture scenes in
several 007 films.

All this practice enhances what is already an impressive
in-built human resilience to perceived threats and
physical damage. The threats Bond must endure are often
fantastical, but people can and do survive all kinds of
life-threatening experiences without extensive training.
Humans would not have survived for so long or been so
successful as a species without some kind of natural
defences.

---

[4] Sodium pentathiol, the so-called 'truth drug', gets a mention in a
few of the films. Khan, in *Octopussy*, says the results are unreliable,
which they are. He has developed his own combination of curare, a
muscle relaxant, and an unnamed psychedelic that Bond says would
result in permanent brain damage. However, it is never used, which
is good because it wouldn't work.

## Positive mental attitude

The brain is central to assessing the world around us and preparing our bodies to respond to perceived danger. If the amygdala, an almond-shaped region deep within the brain, receives information about the outside world it perceives to be threatening, it triggers the release of a cascade of hormones in the body. There are many steps and hormones involved, the most important of which, for our purposes, are adrenaline and cortisol.

Any initial response must happen in a matter of seconds to rapidly shift the body from its normal, long-term, everyday activities to deal with immediate, short-term needs. The release of adrenaline primes the body for the fight-or-flight response within seconds. Pupils dilate to gain as much visual information as possible, resources are diverted away from unnecessary tasks like digestion, and preparations are made for lots of physical activity. The heart speeds up, the lungs breathe more deeply and glucose is released to deliver the raw materials needed for respiration to supply the energy that might be required. Fatty acids are broken down to access more energy stores, and pain sensitivity is dulled. Blood vessels are constricted to prevent bleeding, fibrinogen is produced to help clotting, and blood pressure is raised.

All this preparation uses up a lot of energy, quite apart from the further energy that will be required should an actual threat present itself that needs to be fought or fled from. This high rate of energy consumption is simply not sustainable for long. It also causes damage to the body from reactive molecules released as side products during the respiration process. Initiating this response is vital for survival but it also needs to be shut down as quickly as possible. This is no good if dozens of harpoon-wielding henchmen suddenly hove into view for an extended underwater battle, for example. A second phase of the body's response is required to dampen any negative health effects from the initial response and allow the heightened state to persist for longer if necessary.

Over a period of hours rather than seconds, the hormone cortisol is deployed to rebuild lost energy reserves. Cortisol makes it harder for muscles to absorb glucose and enhances fat storage. It also enhances the immune system response and wound healing – essential if injuries have occurred as a result of the external threat.

Persistent or repeated stress, like life as a double-0 agent, means there will be a lot of cortisol in the body and possibly for longer than is healthy. Too much cortisol means receptors can stop functioning from overuse. Parts of the brain that control the stress response and the immune system can be physically changed as a result.[5] Stress can become chronic and the immune system can be suppressed. Cortisol can change the way the body responds to glucose and stores fat. Too much stress can lead to learning and memory problems, vulnerability to infections, autoimmune diseases, type 2 diabetes, heart problems, and bones can de-mineralise through loss of calcium, leading to osteoporosis.

Bond's brain must be on high alert a lot of the time, seeing as threats to his life occur so often and in so many unexpected ways. Bond's peaceful retirement in Jamaica may in fact be tempered by pain and poor health. Not that he got to enjoy retirement for long

---

[5] These effects can be severe but are reversible.

before being recalled to the service and his ability to function under severe physical duress tested to the full again.

## Pain

Nerve endings in our body can detect heat, cold, pressure and other stimuli. While the biological response of these nerves can be measured, how the brain interprets the messages it receives can't. Perhaps Bond has a naturally high tolerance to pain, physiologically and/or mentally. It would certainly be a very handy trait to have for someone who is injured as regularly as he is. Experiencing pain is an important survival mechanism, as discussed in chapter 019, but our bodies also modify our perception of pain to suit the circumstances.

As already mentioned, part of the fight-or-flight response is to dial down pain perception. If there is a threat that could cause physical harm, then the body needs to keep functioning to fend off that threat and doesn't need to be distracted by pain. This pain tolerance can reach incredible levels. People in battles can continue fighting until physical damage or blood loss becomes so great it prevents their body from functioning. A soldier may be almost unaware of their wounds until the adrenaline rush has abated. Hormones, such as dopamine and serotonin that are part of the body's reward system, are released to help soothe anxiety and reduce the distress the body is experiencing.[6] But behaviour that can

---

[6] There is evidence dopamine and serotonin also help the brain assess risk and make decisions, which would explain a lot about Bond.

increase the activity of dopamine in the brain can become addictive.

Thrill-seekers have been shown to have fewer inhibitory receptors that regulate the dopamine response when compared to the more shy and retiring types. Those who have become accustomed to being on high mental alert can find normal everyday activities, like filing their expenses, writing reports or catching up with friends, as Bond must do between missions, seem particularly mundane and unfulfilling. Seeking out big thrills to fill the dopamine void would explain a lot, like why Bond uses a personal jetpack to escape a French château when it would have been quicker just to run out of the front door (chapter 018). There is also some evidence that risk takers have a harder time learning from their mistakes and negative experiences, which would certainly apply to Bond's repeated willingness to launch himself into dangerous situations.[7]

Another way to tap into the body's reward pathway and increase the availability of dopamine is through the use of chemicals (see chapter 016). In the novels, 007 pops an occasional Benzedrine tablet when he knows he needs to enhance his performance in a stressful situation. Benzedrine, the trade name for methamphetamine, was first sold over the counter as an asthma cure in 1929. In 1936, Benzedrine tablets were given to athletes in the Olympic Games, improving their performance enormously. Dr Fritz Hauschild, a German chemist, was

---

[7] And perhaps also to those of us who have gone back to the cinema 25 times to watch the same events roll out on screen in ever more violent and explosive ways.

inspired by these doped athletes to develop a very pure form of methamphetamine, branded as Pervitin and released on the market in 1937. It proved to be tremendously popular. During the war, Pervitin's abilities to promote aggression and lessen fatigue were put to use by air crews and they became known as 'Stuka tablets'.[8]

While the films have shied away from showing Bond popping pills,[9] his high alcohol and nicotine consumption is there for all to see. It could be just another, more socially acceptable way of self-medicating. Nicotinic receptors, so-called because they are activated by nicotine, are found throughout the body. Activation of nicotinic receptors in the brain causes an initial stimulation, alertness, decreased irritability or aggression, and a reduction of anxiety. At higher doses nicotine becomes a depressant with pain-relieving properties.

Alcohol is another drug that can affect the brain, giving a pleasurable feeling and removing inhibitions, both of which may be helpful to Bond on a mission. It dulls pain but also response times, a side-effect he would want to avoid in situations that require quick thinking and fast action.

Bond's drinking, occasional drug use and smoking (in the books and early films) are understandable as a way to deal with the incredible physical and psychological strain he must be operating under. And, of course, if Bond is going to be a chain-smoking alcoholic, the cigarettes

---

[8] At peak use, during the Blitz, 35 million methamphetamine tablets were issued by the German military in just three months.

[9] In *Skyfall*, Bond's psychological assessment hints at a dependency on alcohol and pills.

and the alcohol he chooses for self-medication would only be the best. He drinks champagne and vodka Martinis, shaken not stirred.[10] The cigarettes are a special blend of three types of Turkish tobacco emblazoned with three gold bands, signifying his rank of commander.

Bond's disastrously bad habits were modelled on Fleming's own daily intake of 70 Morland specials and at least a quarter of a bottle of gin, as he admitted to one doctor he consulted after experiencing chest pains.[11] In Fleming's case it led to his early death, aged only 56, on 12 August 1964.[12] A similar intake should have brought the career of his fictional character to a stumbling, wheezing halt very early on. Over time, Bond has given up the fags and moderated the booze, but his body has gone through plenty of other punishments.

### It's not the years, it's the mileage

In any normal world, Bond should be a wreck of a man, barely able to stand upright let alone chase down criminal masterminds. After so many missions, he should be a bundle of broken bones and torn muscle barely held together by scar tissue (see chapter 019). He should have enough shrapnel rattling around in his body to set off metal detectors. The countless hordes of minions who have taken potshots at Bond can't all have missed. Quite apart from the physical damage done by bullets,

---

[10] 'Three measures of Gordon's, one of vodka, half a measure of Kina Lillet. Shake well until it's ice cold, then add a large thin slice of lemon peel.'

[11] Later, another doctor he consulted learned he had managed to cut back his cigarette intake to 60 a day.

[12] Coincidentally his son's 12th birthday.

there could also be slow and insidious effects on his health due to lead poisoning.

With modern medical treatment, the chances of surviving being shot are actually very good (around 70 per cent of 115,000 annual firearm incidents in the USA are not fatal). Unless a bullet, or bits of it, are immediately threatening someone's life, they are usually left inside the body. Over time, the lead most bullets are made with has the potential to seep into the body and cause widespread problems, as a study from 2017 highlighted. It damages nerves, kidneys and, because it is chemically similar to calcium, it can be stored for years in the bones. Bond may well be struggling through his missions while feeling quite unwell, with potentially hypertension, poor kidney function and tremors that could seriously impair his marksmanship. The effects lead can have on nerves might also explain Bond's occasionally erratic behaviour.

Some Bond bad guys, however, are not content with conventional ammunition. Scaramanga's gold bullet would have had little impact, at least chemically (see chapter 009), on Bond's body, but Scaramanga is killed before we have the chance to find out. The same cannot be said of the depleted uranium bullet used by Patrice to shoot Bond in *Skyfall*.

Depleted uranium is much less radioactive than uranium ore, because the most radioactive atoms have been separated from it to be used in nuclear reactors. It is therefore the chemical, rather than the radioactive, consequences of interaction with this metal that is of concern. Uranium preferentially distributes to bones, the liver and the kidneys. Bone deposits last about 1.5 years, but it takes only days for the body to excrete it

from other areas. The most serious effects of exposure
to depleted uranium are impaired kidney function.
Bone growth is also diminished and can result in
osteoporosis.[13] Depleted uranium is used in tank armour
and in missiles because of its high strength, so how a
bullet made from it would fragment inside Bond's body
is difficult to understand. It should have passed straight
through him, leaving behind considerable damage but
not shards of metal. Nevertheless, we see Bond digging
out some scraps of bullet from his shoulder. Let's hope
he removed it all.

Being shot hurts, but Patrice's aim was to kill Bond.
Le Chiffre wants Bond to experience as much pain as
possible but without killing him. Having prepared
ourselves mentally, physically and chemically, it's time to
tackle the subject we have been avoiding for most of this
chapter. Cross your legs, here we go…

**The knotted rope**
In 007's violent world there are some rules, or at least
gentlemanly etiquette, around behaviour. Bond kills and
maims, but only the bad guys, and he never resorts to
torture. The villains, on the other hand, revel in
humiliating or physically threatening Bond and others.
Few stoop so low as Le Chiffre in *Casino Royale* who,
quite literally, hits below the belt. If there is a case for the
defence of Le Chiffre's actions, it is that Bond has gone
to considerable lengths to provoke him.

---

[13] There was concern after soldiers from the Gulf War were left
with depleted uranium fragments in their body and potential long-
term effects, but as of 2013 there had been no studies on the case.

Le Chiffre acts as a kind of private bank for criminals. He holds their money, making some canny investments while he has it, and returns it as and when they have a big new plan they want to put into operation. Thanks to Bond's intervention (stopping a new passenger plane being blown up), Le Chiffre loses a lot of money and plans to recoup it by setting up a high-stakes poker game. M sees an opportunity to permanently dry up funding for major criminal activity by getting Bond to win the game – MI6 doesn't have to get its hands bloody by assassinating Le Chiffre because his disgruntled clients will do it for them.[14] Bond wins, of course. But Le Chiffre still needs the money, so he captures Bond and tortures him to extract the password for the account containing all the winnings.

The film took its plot directly from Ian Fleming's novel of the same name. Fleming had been inspired by a trip to a Lisbon casino during the war where, in a dimly lit, uninspiring grey building, he lost money to Portuguese businessmen in a low-stakes game.[15] But Fleming said to his boss, who was with him on the trip, 'What if those men had been German secret service agents and suppose we had cleaned them out of their money; now that would have been exciting.' His fantasy version of events became the basis of his first novel. Relatively few changes were made for the 2006 film adaptation. The card game is updated from baccarat to

---

[14] And the winnings will no doubt come in handy paying for those expensive cars and gadgets Bond is going to wreck in the process.

[15] Real-life spy Dusko Popov, who also spent time in Lisbon's casinos during the war, recounted a different version of events (see chapter 024).

poker, and a carpet beater is exchanged for a knotted rope[16] in the infamous torture scene.

To get his information as quickly as possible, Le Chiffre obviously wants to maximise the pain experienced by Bond without killing him, and so targets a very sensitive part of Bond's anatomy. In fact, it's the eyes that have the most nerve endings, but another densely enervated part of the body is chosen as the target. The knotted rope may exert more force than a carpet beater but it will also need more skill in finding its target(s) – suggesting Le Chiffre has practised.

Anticipating pain can make it feel worse when it arrives; perhaps this is why Bond villains spend so long describing what they are about to do to 007 when they capture him. But there are some things Bond can do to reduce the pain, even when tied to a chair. In the novel, Bond only has to suffer one hit and he does this uttering no more than a low groan. Bond of the film has a more prolonged experience and is much more vocal about it. Vocalising, like saying 'ow' or swearing,[17] has been shown to increase a person's ability to endure pain. Perhaps this is part of the reason Bond is always keen to answer his attacker's physical abuse with quips. Physical distraction, like button pressing or rocking back and forth on the chair you've been tied to, can also help, at least a little bit. Some people might even enjoy the experience. There are those who would pay good money to be in Bond's position, including Fleming if the way

---

[16] Apparently known as a 'monkey's fist'.

[17] Swearing is best, especially if you don't swear much in everyday life.

he lovingly describes the sensations of torture in his novels are anything to go by.[18]

There are so many nerve endings in the genitals for a reason. Without the pleasurable experiences they can trigger, our species would have died out long ago. The boundary between pain and pleasure is up to interpretation in the brain and will differ from one person to another.

Unprotected by bone or cartilage, the soft tissue of the testes is particularly sensitive to pressure that triggers the nerves within them. It also means they are more susceptible to damage and, though Bond is saved from his excruciating experience, he still needs medical treatment afterwards. There appears to be no lasting damage, and Bond is back on the job in no time.

---

[18] No judgement intended. Whatever works for you. Please stay safe.

# 022

# *Quantum of Solace* and the girl covered in oil

The 007 franchise is extremely body-conscious. The films and books have become renowned for them – shapely ones, strong ones, scarred ones, but mostly dead ones. The high body-count in every film is mostly made up of dead minions left in crumpled heaps on the floor or as scattered remains somewhere off camera.[1] Explosions and gunshot wounds are the main, but by no means only, ways to die in a Bond film. Variety is the spice of life and, in Bond's world, it's also the spice of death.

So that these incredibly violent films can retain family-friendly film certifications, the blood and the gore is kept to a bare minimum. Sometimes the camera lingers over a corpse for confirmation that key figures have received their just deserts. Occasionally, the death may be so unusual that the audience needs a little time to process what is happening. A few deaths have been devised for their visual spectacle. And there is one death that looks so good it has been put on posters and magazine covers.

---

[1] Accurate kill counts are difficult to make without complete staff lists for all those lairs that get blown up. Even just counting the bodies you see on screen isn't straightforward when most of them are dressed in identical minion boiler suits or military uniforms. But several people have tried and the total is well over 2,000.

The character of Jill Masterson covered in gold paint has become one of the defining images not just in the Bond franchise but in cinema history. The image is so striking, it has been reproduced and reinterpreted many times over.[2] The 2008 Bond film *Quantum of Solace* paid many visual homages to previous films from the franchise, but Agent Field's dead body lying on a bed covered in crude oil is easily the most identifiable and memorable. The pose and the setting are identical to Masterson's body in *Goldfinger*, only the material covering her has been changed from the 1964 original.

## Breathtaking

*Quantum of Solace* has Bond on the trail of Dominic Greene, head of a mysterious criminal organisation called Quantum. Greene presents himself to the world as an environmentalist, though he is anything but. He is negotiating the purchase of a seemingly worthless expanse of Bolivian desert. The CIA, the Bolivians and MI6 believe he is really after oil, and to maintain this fiction, Greene deliberately leaves some very crude clues.

When Bond arrives in Bolivia, he teams up with the local MI6 operative, Strawberry Fields,[3] but not for long. To remove one potential problem, and hammer home the supposed oil connection, Greene has Fields killed. Her body is found lying prone on the bed in her hotel room, covered head to foot in crude oil. There is

---

[2] Is it weird that the image of a dead woman has become so idolised? I think it's weird.

[3] Yes, that really is the character's name.

yet more oil in a pool below her head where it has drained out of her body. The oil in her lungs prevented oxygen reaching the rest of her body. Fields has been drowned.

Bond, rarely one for sentiment, is more concerned that the oil is a deliberate red herring and Greene is in fact after the water he has dammed up under the desert. The plot moves on and Fields is forgotten, at least within the context of the film. The audience probably remembered the image of her dead body longer than Bond did, and not just because it is an unmistakable tribute to the girl covered in gold in *Goldfinger* from 44 years earlier.

## Setting the gold standard

The idea of a golden girl came from Ian Fleming, though quite where he plucked it from is a mystery. One internet fan site claims he was inspired by a Swiss model who was accidentally covered in paint, but no such case has been found. Another suggestion was that he might have read *The Romance of Leonardo Da Vinci* by Dmitry Merezhkovsky, where a woman experiences a similar fate, but no one knows for certain.[4] And though the image of a woman naked except for a thin layer of gold paint seems made with its cinematic impact in mind, Fleming wrote about it in 1958, long before any film deals had been signed. What is more surprising is that what has become a very sexualised

---

[4] Another remote possibility is a short story by Dorothy L. Sayers in 1928 featuring her popular detective Lord Peter Wimsey and a woman murdered by being electroplated.

image is not presented in quite the same way in the novel.[5] The strange death of Jill Masterson is reported second-hand and in a very matter-of-fact way. The method seems to have intrigued Fleming more than how it looked.

Bond learns what happened to Jill from her sister, Tilly. She explains that every month Goldfinger would have one of his staff paint a woman gold, all except their backbone because, 'If their bodies were completely covered with gold paint, the pores of the skin wouldn't be able to breathe. Then they'd die.' Jill became one of these golden women, only her backbone *was* painted. She phoned Tilly from the emergency ward in a hospital, where she died that same night. Tilly then contacted a skin specialist who tells her about a similar case: a cabaret dancer who died after being painted silver to pose as a statue.

The film version is a little different. Bond is hit on the back of the head by Oddjob and Jill is painted while he is unconscious. He wakes up to find her dead body. The painting process and the death happen much faster, but the same story about the cabaret dancer is repeated. The detail of the cause of death – that unless a small patch of skin is left uncovered the paint blocks the pores and the skin can't breathe – is also repeated. The image of Masterson's gold body lying on the hotel bed has become ingrained in popular culture and the idea of being killed by skin suffocation has become accepted as fact, even though it is nonsense.

---

[5] The image is still sexual, but it is presented as an unusual one, and a sign of Goldfinger's perversion.

It is true that some oxygen is absorbed through the skin, but it is only around 2 per cent. Cutting off such a tiny fraction of the body's oxygen supply is not going to make a huge difference to a person's health. Cutting off the 98 per cent that is absorbed through the lungs is a much more serious matter, as illustrated by what happened to Agent Fields in *Quantum of Solace*. However, skin has many other important functions that can affect well-being.

The pores of the skin are primarily for excreting substances rather than absorbing them. Sweat is exuded through the pores so that waste material can be removed from the body but, more importantly, so that liquid evaporating from the surface of the skin cools it down. Blocking the pores with paint could prevent sweating, making it difficult for the body to regulate its internal temperature. Masterson's naked gold body probably got a few people hot under the collar, but what about Masterson herself?

Normal body temperature is around 37°C (98°F), well above normal room temperature, and so our bodies constantly generate heat, some of which is lost to our surroundings. Someone of Jill Masterson's petite size is probably radiating around 70 watts (W) of power to keep her warm. If those 70W were to remain trapped in her body, however, she would heat up very quickly, because it only takes a small increase in body temperature for things to go very wrong. A fever of 42°C (108°F) can be fatal because cells are programmed to self-destruct at this temperature in case they are infected. The longer the body is kept above this temperature, the more cells will die. All cells are important but some cells are more important than others. If too many brain cells die so will the person because these cells control vital functions such as breathing and heartbeat. But can a layer of gold paint really make that much difference?

Physicists Tolan and Stolze were curious to find out. They calculated that 42 of Masterson's 70W would be lost through her sweat evaporating.[6] Assuming every pore on her body is blocked, the heat retained would result in her death around six hours after she was painted. Bond and Masterson must have been unconscious for at least that amount of time, because if either of them had woken up and noticed the overheating, a cold shower could have saved her.[7] Even if the paint didn't wash off in the water (in the novel it is said a special resin is needed), there are plenty of ways to cool a body until a suitable solvent could be found.

Alternatively, death might be due to the paint itself rather than the fact it blocks Jill's pores. If Goldfinger uses real powdered gold in his paint – and I can't imagine, given his obsession with the metal, he would accept any substitutes – Jill has nothing to worry about, at least chemically. Gold is a very unreactive metal; it's why it's a popular choice for jewellery because it won't tarnish or irritate the skin even after prolonged wear. But gold is also expensive, so cheaper substitutes are often used and this might help explain Masterson's death.

---

[6] The other 28W will be dissipated to the surroundings. The layer of paint would have to be much thicker to prevent this. The shiny finish won't reflect the heat back either, because it is in direct contact with the skin.

[7] Cold showers could have saved Bond from a lot of bother over the years.

'Dutch metal' or 'Dutch gold' is sometimes used to make imitation gold leaf, which is actually an alloy of copper and zinc, both of which combine to give a yellowy colour similar to gold. These metals are more reactive than gold and both have known biological functions. Copper can trigger allergic reactions, leaving a rash, itching and small blisters on the skin. Someone who is particularly sensitive to copper could have a severe reaction to it, especially if it was used to cover their entire body.[8]

In Fleming's novel, there is another clue as to what might have happened. Masterson is painted gold, but the cabaret dancer she is compared with was painted silver. It may not seem like much but it could make all the difference. Silver is more reactive than gold, but the human body is extremely tolerant of the metal.[9] However, the colour and appearance of silver is also similar to a lot of other metals, meaning real silver is unlikely to be used in body paint when so many other cheaper alternatives are available. And other silver-coloured metals are not always accommodated so easily by the human body, as illustrated by the unfortunate experiences of the actor Buddy Ebson.

In the 1939 film *The Wizard of Oz*, the Tin Man character had to appear as though he were made of

---

[8] Leaving a small patch of unpainted skin would hardly help if the rest were undergoing an allergic reaction.

[9] Silver poisoning can certainly happen, but it is very unlikely to kill someone. Argyria is a condition caused by excessive exposure to silver. The main symptom is that the skin turns blue permanently, though there may also be decreased kidney function and loss of night vision.

metal. Ebson had his hands, neck and face painted in a layer of white make-up, which was then sprayed over with aluminium dust. As filming went on, Ebson started to suffer from shortness of breath and cramping in his toes and fingers. He was experiencing the first signs of an allergic reaction to the aluminium. At the end of his first week on set he woke up in the middle of the night with severe muscle cramps that travelled across his body and caused his lungs to collapse. He spent two weeks in hospital and the role of the Tin Man went to another actor, Jack Haley.

When Fleming thought about someone being hospitalised as a result of being covered in paint, perhaps he was thinking of what happened to Ebson. The author simply swapped silver for gold to fit with the novel's theme, not realising it wouldn't have the same effect. Fleming's explanation of suffocation sounds like a muddled version of Ebson's collapsed lungs, and the novel moves on before the reader has time to think about it too much anyway. The same applies to the filmmakers, who wanted a stunning image, not a documentary on the science of skin or metal poisoning. None of this means covering Masterson in paint wouldn't kill her, just not in the way it is explained in *Goldfinger*.

For the actress who was painted gold to film the scene, she was in no real danger.[10] Shirley Eaton's biggest complaint was not about having the paint applied, but getting it off afterwards. 'They scrubbed me down until I

---

[10] Nevertheless, a strip of skin on her front was left unpainted, just in case.

was pink raw. And a week later I sweated it out in a Turkish bath.' She may be the most remembered, but Eaton is far from the only actor to be made uncomfortable for the benefit of a visually arresting Bond death.[11]

## Some like it hot

Jill Masterson's demise may have been due to a subtle change in body temperature,[12] but other deaths have left no doubt that temperature was a factor, because the character literally goes up in flames. First was the unfortunate Quarrel in *Dr No*. In a confrontation with Dr No's not-so-terrifying dragon tank (see chapter 018), Quarrel is burned by the flamethrowers bolted to its snout/bumper and left for dead, although he is probably still alive.

There have been many weapons that could be described as flamethrowers stretching back to the fifth century BC. The flame-throwing dragon tank shown in *Dr No* was probably based on the Second World War design by Richard Fiedler. The Kleinflammenwerfer used petrol, tar and pressurised air to shoot flames almost 20m (65ft). The devices were small enough to be carried by a person but they could also be fitted to tanks.

---

[11] The experience of filming it was no more comfortable for Gemma Arterton who played Agent Fields: 'I couldn't move, I couldn't see, I couldn't breathe or hear because the oil went into my ears. It was unpleasant, but it's something I'll always remember and it will be an iconic part of the film.'

[12] Had Fleming or the filmmakers known this they would undoubtedly have made some pun about smouldering looks or being too hot to handle.

There is no doubt that a flamethrower can cause severe burns, as the scientist James Lovelock can testify. He was involved in studying the effects of heat radiation during the Second World War and said, 'I shall not forget the fierce radiant heat from the wall of flame projected by one of these weapons.' He and a colleague were ordered to test the effects of heat on shaved rabbits, but both baulked at the idea of burning even anaesthetised rabbits. So they decided to experiment on themselves. He described the feeling of being burned as 'exquisitely painful' but, to his surprise, after about a week of experiments, the pain subsided into a sensation of pressure. Endorphins flowing through their bodies were acting as natural analgesics.

At 44°C (111°F), it takes five hours for a burn to appear on the skin. Above 60°C (140°F), it only takes three seconds. The higher the temperature, the faster the burn. The temperature of the flames coming out of a Kleinflammenwerfer would be much, much higher. Even if Quarrel managed to put out the flames quickly, and there is plenty of water in his swampy surroundings to help, the damage to the skin would be rapid and severe. But it might not kill him, at least not immediately.

The body does everything it can to preserve life. The most important cells are those in the major organs inside the trunk of the body and the skull, where they are protected from physical damage and heat by being encased in bone and/or fat to insulate them. Some cells, like skin cells, can be sacrificed to preserve others, but severe burns can still kill. Our skin shields us from external harm, but it also holds us together. Without our waterproof covering we literally evaporate. Burns victims

who have lost much of their skin surface can require up
to 20l (4.4gal) of water a day just to stay alive.

Further problems can come from gases inhaled in
fires. In a situation like a house fire, it is most likely that
toxic gases from the flames, or lack of oxygen, kills
people before the flames do. Quarrel is outside and will
have plenty of oxygen available to breathe. Any toxic
gases from the burning tar would also be dispersed in the
open air. However, breathing in hot gases can easily
damage the delicate lining of the throat and lungs. This
would prevent Quarrel from being able to absorb oxygen
into his body efficiently. He would likely succumb to the
combined effects of shock (the heart and lungs going
into overdrive to try and preserve life as long as possible),
damage to the lungs from inhaling hot gases, and
dehydration. It would be an excruciating and prolonged
death.

All we see in the film are flames shooting towards the
bush Quarrel is hiding behind and we hear his screams.
We're not shown Quarrel burning, or his remains,
because he is one of the good guys. Villains, on the other
hand, are visual fair game when it comes to being
immolated. Mr Wint tries to skewer Bond with some
flaming kebabs in *Diamonds Are Forever*, but Bond douses
him in brandy so the flames spread over his body.
Stamper, in *Tomorrow Never Dies*, is burned by the exhaust
from a stolen missile. And Franz Sanchez in *Licence to Kill*
becomes a 6ft flaming torch when he is soaked in his own
cocaine-enriched petroleum and set on fire. The stunt
actor, Paul Weston, taking the actor Robert Davi's place,
was dressed in two firesuits, covered in ice-cold anti-
flame gel and had to use bottled oxygen to breathe. He

had three minutes of breathable air but half that time was used up getting into the protective gear. For the remaining minute and a half of filming he had to walk a carefully rehearsed path to where people were waiting to extinguish him.

Flames are a common sight in a Bond film. But the other end of the temperature range isn't forgotten, and that's where we find Boris. He provides the technical support for the bad guy's master plan in *GoldenEye* and meets his end when tanks of liquid nitrogen explode, freezing him in an instant.[13] His corpse is prominently displayed, in triumphant pose and with icicles dripping from his arms and face, for comic effect.

## Taking things to extremes

All life, so far as we know, relies on water being able to flow. It is the medium that transports elements and molecules from one place to another and allows the billions of complex chemical processes to be carried out that are necessary for life. Freezing water into solid ice halts these processes but, because of the way water freezes, life can't always pick up from where it left off once things start to thaw.

To become a liquid, nitrogen must be cooled to a temperature of -196°C (-321°F). The damage from having gallons of it pour over you will be severe, though probably not quite enough to turn someone into the human popsicle we see on screen. As the temperature

---

[13] Why the bad guy's lair needs so much liquid nitrogen is never explained, or why it is stored right next to excessive amounts of fuel.

drops, protein molecules, which must maintain a certain shape in order to work properly, begin to unwind. As well as unravelling, things also start to leak. Cell walls are largely made up of fat molecules and, as the thermometer falls, these fats will start to congeal and clump together, leaving gaps where the cell contents can leak out.

If the water outside the cells starts to freeze, liquid water will be sucked out from inside the cells, dehydrating them.[14] Ice crystals forming inside or outside cells can puncture cell walls with their sharp edges and points. Water also expands when it freezes, meaning cells can be ripped apart. No cell will survive if its internal water freezes. This is why frostbite causes fingers to turn black and why fruit and veg often seems mushy after being defrosted from the freezer.

One way to prevent cells rupturing is to freeze things very quickly, so the molecules of water don't have time to arrange themselves into their expanded conformation. Liquid nitrogen is so cold it will freeze water very quickly, but this will be of little comfort to Boris. His skin might freeze solid almost instantly, but his internal body heat will keep his internal organs warm a little longer, giving time for those damaging ice crystals to form. Even if Boris could be frozen solid in an instant, at liquid nitrogen temperatures the body's organs are susceptible to cracking. A sharp knock and Boris will start to fracture.

Body heat can cause other problems too. In *The Man with the Golden Gun*, Scaramanga is keeping important bits of his solar power station cold with liquid helium, at

---

[14] This is the principle behind freeze-drying.

a temperature of -269°C (-452°F). When the technician falls into one of these liquid helium pools, his body heat starts to warm the helium and starts the process that results in the whole island lair blowing up.[15] At such low temperatures, open vats of liquid helium can start to condense oxygen out of the air into a liquid. Liquid oxygen is highly concentrated and can fuel some very nasty fires. Install a lid or some safety rails and *The Man with the Golden Gun* could have ended very differently.

## Cause of death

No matter how unusual the method, every single death comes down to the same basic cause – a disruption to the body's oxygen cycle, and the breakdown can be at any point. The most obvious break might be preventing oxygen getting into the body in the first place, for example Agent Fields' oil-filled lungs[16] or Quarrel's heat-damaged lungs.

Stopping oxygen circulating so it can't reach vital organs will also work. Take the example of being broken up into little pieces, a fate that befalls a surprising number of people in Bond's world. Whether the disassembling is done by a snowblower (*On Her Majesty's Secret Service*), cocaine crusher (*Licence to Kill*) or sea drill (*Tomorrow Never Dies*), the result is the same: oxygen cannot be

---

[15] The technician would be completely submerged and in much colder liquid than Boris experiences, but even if there had been someone around to fish him out and warm him up, the prognosis is not good.

[16] Also Plenty O'Toole who drowns in a pool, Vesper Lynd who drowns in a lift, and the dozens of mariners who go down with the many ships and submarines that have been sunk in the franchise.

delivered around the body because the pieces are no longer connected. Blood loss from bullet or stab wounds means blood is leaking out all over the place rather than pulsing through the network of arteries and capillaries as it should be.

Finally, the processing of oxygen in the body to release energy can also be disrupted. Cells destroyed by heat or broken apart by ice can't use oxygen. Cyanide, the secret agents' favourite poison, stops oxygen binding to enzymes involved in the respiration process. But that is the subject of the next chapter.

# 023

# *Skyfall* and the cyanide capsule

*Skyfall* is a back-to-basics Bond film. It is all about personal vendettas rather than international threats, and Bond is forced to rely more on his own resources than the sophisticated technology and infrastructure of MI6. Consequently, the film features minimal gadgets but includes plenty of other classic elements popular from previous incarnations of the franchise. There is the original DB5,[1] and Q and Moneypenny make their return after a two-film and 12-year absence. With the exception of the car, which is a perfect replica of the *Goldfinger* original, all these traditional elements are given a new twist. For example, Q is younger than Bond, and Moneypenny is shown as a field agent before taking her desk job. Another traditional element is an addition from the world of classic spy fiction – the secret agent's standard-issue cyanide pill, but again with a twist.

Cyanide is associated with spies and espionage more than any other poison. It can be no surprise that it features in many 007 films and novels. No explanation is given as to what the effects of cyanide will be or how long those exposed to cyanide can expect to live. Everyone knows that cyanide will kill you, and very quickly. That was until *Skyfall* was released.

---

[1] Quite why a 2012 James Bond has a 1960s spy car complete with working 1960s gadgets in his private possession is never explained. But who cares? It's a great addition to the film.

In the 2012 Bond film, the villain, the delightfully unsettling Raoul Silva, is revealed to be a former MI6 agent, so would be expected to have a cyanide pill in his possession. But when he came to use the pill, it did not work as he hoped. It evidently didn't kill him because he is shown, very much alive, and describing the agonizing, drawn-out effects the poison had on him.

Over the course of many 007 capers, cyanide has been shown to have a range of effects, from keeling over dead, but otherwise undamaged, in a few seconds, to being very much alive but horribly disfigured. So which is it? Films, especially Bond films, are not known for their strict adherence to scientific facts or narrative logic but, as we have seen, not all of it is made up. As a chemist, I have a particular interest in cyanide and the science around it, so indulge me in a more lingering look at this iconic chemical.[2]

## The science of cyanide
Cyanide is a simple chemical unit, or functional group, made up of one carbon atom and one nitrogen atom. This group forms part of a larger molecule, the toxicity of which depends on how easily the cyanide unit can be broken off. Hydrogen cyanide and cyanide salts, such as potassium or sodium cyanide, are exceptionally toxic because the cyanide unit is easily released.[3] Once freed,

---

[2] More lingering than most victims of cyanide poisoning in Bond films.

[3] Some cyanide compounds hold on to the cyanide unit so tightly they are perfectly safe to eat.

the damage cyanide can do to a living thing is very specific, but it can affect every part of the body.

Cyanide kills because it cripples the body's energy supplies. Released from its parent compound, the cyanide unit is rapidly absorbed into the bloodstream where it is carried to cells around the body. Most cells contain mitochondria, the biological engines that combine oxygen and glucose to release energy in the process of respiration. Without a constant supply of energy, cells die. If enough cells die, the organ fails. If several organs fail, the result is death.

Respiration involves a series of chemical steps, each controlled by an enzyme. Cyanide stops the whole process in its tracks by attaching to just one of the enzymes in the sequence, cytochrome oxidase. It blocks the site where an oxygen molecule would normally bind. Therefore, no matter how much oxygen is available, the enzyme simply cannot process it to release energy. Cells that need the most energy, like nerve cells, are most likely to feel the effects first. Damage to nerve cells can cause disorientation, blinding headaches, convulsions and coma before death. However, experiences vary. Sometimes a person is rendered unconscious before symptoms can manifest themselves. Sometimes all these agonising symptoms can occur over a period of minutes before cyanide finally kills. Though it is a very narrow window of time for medical intervention to try and counteract the poison, it must surely feel a lot longer for the person experiencing the symptoms.

A poison that causes death in less than 10 minutes is considered to be very fast acting. Many other poisons take a lot longer to kill – sometimes hours, days or even weeks. If a secret agent is captured, they need to be dead very quickly, before torture and other forms of persuasion cause them to leak their secrets to the enemy. The toxin of choice also needs to be lethal in small quantities so that it can be carried discreetly. Cyanide has the added advantages that it is easy to produce and has a long history of use, in lethal and non-lethal applications,[4] creating a large body of information scientists can draw on. It was therefore the obvious choice to use sodium or potassium cyanide in the 'L-pills' that were issued to individuals carrying out top-secret missions during the Second World War.[5] The L-pills were made of glass that was easy to break but wouldn't allow any cyanide to seep out until it was needed. But they would also have been fragile and not the sort of thing that would survive the many scrapes Bond regularly finds himself in.

Specific antidotes to cyanide poisoning have been developed because of its widespread conventional use. There are many formulations and variations but most work on the premise of offering an alternative, and more attractive, site for the cyanide to bind to before it can get to the cytochrome oxidase and cause problems. The key is to administer it in time. Any interrogation officer worth their salt would have a cyanide kit to hand if they were about to question an enemy agent. A

---

[4] Industrial applications such as metal refining, the chemical industry, and even some food preservatives rely on cyanide-based compounds.

[5] At the end of the Second World War several senior Nazis took their own life using cyanide, including Hitler. Though he swallowed cyanide, he felt it was taking too long and so he shot himself.

twenty-first-century spy might therefore be expected to carry an alternative poison in their suicide pill. Even as early as 1960 the US was using toxic alternatives. The U-2 spy-plane pilot, Gary Powers, was issued with shellfish poison[6] secreted in the grooves of a tiny drill bit that was in turn hidden inside a silver dollar.[7]

These new poisons were being issued a few years before the first Bond film was made, but obviously the Americans were not about to tell filmmakers about their latest developments in espionage technology so they could broadcast it around the world. And, while the team behind the Bond films always wants to include the latest science and gadgets, there is a balance between being at the cutting edge and established enough for audiences to be familiar with it. It is understandable that the franchise has stuck with cyanide for so long, even if its use has been updated over the years.

## Classic versus cliché

In the Sean Connery era of 007, the cyanide pill was standard. In *Dr No*, Bond arrives in Jamaica and almost immediately runs into one of the bad guys who he beats up to get information. The bad guy promises to tell all but asks for a cigarette first and Bond obligingly hands him his pack. The henchman collapses on the ground dead, seconds after biting into one of his cigarettes. Bond sniffs gingerly at the cigarette packet

---

[6] Possibly something like the tetrodotoxin poison discussed in chapter 002.

[7] When Powers was captured he decided not to take the poison. KGB agents found it and tested it on a dog. The dog was dead within 10 seconds.

and knows it is cyanide before any lab tests have been carried out to confirm it.[8] No explanation about how he reached this conclusion is given because most people know that cyanide kills very quickly and has the characteristic smell of bitter almonds. In fact, bitter almonds smell of cyanide because they contain a lot of amygdalin, a compound with a cyanide unit that is relatively easy to break off.

Bond shows the tainted cigarette to his contact at Government House in Jamaica, who is apparently impressed with the devious nature of hiding the cyanide capsule inside a cigarette. This seems quite naïve on the contact's part. Well-known murders in fact and fiction have involved cyanide added to all sorts of everyday items including drinks and chewing gum. It is more likely that he is surprised at how serious the mission is, given that someone has killed himself within minutes of Bond landing on the island rather than give away any secrets of the evil plan.

Even less explanation is presented when a cyanide pill is used in *Thunderball*. MI6 agent Paula has been captured but when a couple of henchmen go to question her they find her dead. There is a capsule in her mouth and they conclude it was cyanide, wisely without even a sniff to confirm their theory. Paula and the cyanide pill are quickly forgotten. The poison didn't make another appearance in the franchise until 12 years later, and under a rather different guise.

---

[8] Be careful before you sniff anything. Around 40 per cent of the population can't smell cyanide. The reason is believed to be a mix of genetic and environmental factors.

The 1970s saw a new Bond, Roger Moore, and new ways to weaponise cyanide. A key moment in the 1977 film *The Spy Who Loved Me* is the capture of a US submarine by Karl Stromberg as part of his plan to start a world war so he can be left in peace in his underwater lair. Swallowed up inside a giant supertanker, the submarine crew nervously wait to see what happens next. Huge metal bolts are fired into the side of the vessel, penetrating the hull, and a hose is put in place to connect the bolts to gas cylinders. A speaker is also attached to the hull so the crew can hear Stromberg's threat loud and clear. Unless they surrender, the crew face extermination by cyanide gas. No one doubts Stromberg's hull-piercing cyanide spray guns will work, and the crew emerge from the submarine with their hands in the air. It's a sensible move.

Penetrating a submarine hull is not trivial, but certainly within the capabilities of a villain who has the resources to build a supertanker and an underwater laboratory/lair. Hooking the bolt up to a gas cylinder is a low-tech but practical way to introduce the cyanide. A similar but more sophisticated real-life weapon was developed during the Second World War.

A UK team conducting research into chemical and biological warfare (see chapter 006) built a cyanide-containing anti-tank projectile. The weapon could pierce a hole through a tank's armour using a hollow charge.[9] Liquid hydrogen cyanide could then be squirted inside to kill the crew without having to stop the tank to hook

---

[9] A cavity inside the weapon directs the force of the explosion to create a greater impact that can pierce armour plate.

it up to gas cylinders. Other cyanide-laced weapons, possibly inspired by Second World War technology, appeared in another Bond film just two years later.

In *Moonraker*, Q-branch issues Bond with 10 darts that can be fitted to his watch and fired with a flick of the wrist. The gadget is ably demonstrated by Bond as he fires a dart into the flanks of a horse in one of M's office paintings. Q merely rolls his eyes at Bond's tomfoolery and goes on to explain why he should take the weapon more seriously: five of the darts' tips have been coated with cyanide, causing death in 30 seconds.[10]

Darts fired from a watch and triggered by nerve impulses in the wrist might be a relatively new invention for 1979, but adding cyanide to weapons is old hat, as we have seen. As well as the anti-tank weapons developed in the Second World War, there were also small-scale cyanide weapons. The Germans made bullets containing cyanide to increase the ammunition's lethality, though the bullets, just like the anti-tank device, were never used. Ian Fleming liked to boast about his fountain pen that could spray tear gas or cyanide for 'really dangerous missions'.

Decades after *Moonraker*, the Bond films were still going strong and still making use of the formula that had made the franchise such a success. In *Die Another Day*, the Pierce Brosnan Bond film of 2000, M asks why Bond didn't use his cyanide pill rather than face torture, and he replies he threw it away years ago. Obviously Bond can be trusted not to reveal any secrets, even under

---

[10] The other five have armour-piercing tips.

torture (see chapter 021), and it is implied that cyanide pills are old-fashioned anyway. The reappearance of the cyanide capsule in 2012's *Skyfall* is, then, a little surprising, but it is certainly in keeping with the retro feel of much of the film. To ensure the audience is kept interested, this well-worn spy trope does something a bit different.

Every 007 film requires a tense confrontation between Bond and the bad guy in which motives are revealed and plots hinted at, or sometimes explained in great detail, usually over a nice meal or drinks. In *Skyfall*, this moment arrives when Silva is captured and imprisoned in MI6's temporary London headquarters.[11] Silva, the former MI6 agent who has gone to the bad, is reunited with his old boss, M. He relates how his employment with the service was brought to an unpleasant end when he was captured and tortured by the Chinese. There was no attempt to negotiate his freedom or rescue him. Silva felt abandoned by MI6 and M in particular. With no hope left, he decided to use the cyanide capsule that was hidden inside one of his teeth.

The important thing a cyanide capsule has to do is release the poison effectively when required, and only when required. It must be tough enough not to break accidentally but easy enough to break when needed, unlike the L-pills mentioned earlier. In *Skyfall*, Silva states his pill was kept out of harm's way, hidden inside his tooth. I would question the wisdom of this plan. Wouldn't breaking the tooth also break the capsule?

---

[11] The high-class catering we have come to expect from these encounters is notably absent on this occasion. I assume MI6's budget is taken up with other things.

What if you broke your tooth by accident? Could you potentially spot spies by their fastidious attention to dental health and avoidance of crunchy food?[12] I have many questions, but we need to move on.

Whatever it is made of, Silva's suicide pill could easily contain a lethal dose of cyanide and still be small enough to fit inside one of his molars.[13] Silva goes on to give us a lot more information about cyanide pills than any other Bond film has ever bothered with. He talks about hydrogen cyanide and, though it is extraordinarily toxic, it would be a poor choice to put inside a pill. Pure hydrogen cyanide is a liquid that boils at just 26°C (79°F), well below body temperature, and potentially problematic if you are carrying a pill of it about your person. You wouldn't want the capsule to swell up and burst when it got a bit warm. Instead, a cyanide salt, traditionally sodium or potassium, would be the preferred option because they are stable solids but still lethal in small amounts. However, Silva is not wrong to talk about hydrogen cyanide, as we shall see later.

Having extracted and bitten into the capsule, Silva says the cyanide 'burned all of my insides'. But, unlike the henchmen and agents that have gone before him, he did not die. M is unmoved and goes to leave, but Silva calls her back by asking if she knows what hydrogen cyanide does. She turns to see Silva pull a prosthetic out of his mouth that has been occupying the space where the top row of teeth and gums are normally found. What

---

[12] Is this why Silva's meeting with Bond and M isn't catered?
[13] 200–300mg of pure potassium or sodium cyanide will do the trick, in case you were wondering.

teeth he has left look to be in a very poor state and the
skin of his cheek sinks into the gap left by the prosthetic.
The skin itself seems to be intact and shows no sign of
being a graft. Is that what really happens when you
crunch cyanide between your teeth?

The short answer is no. When the capsule broke in the
mouth, the slightly acidic and watery environment found
there would have started converting the cyanide salt into
hydrogen cyanide. Hydrogen cyanide dissolves in water
to make hydrocyanic acid, increasing the acidity but not
by very much. Hydrocyanic acid is a weak acid, weaker
than the acetic acid in vinegar, though the acid would be
concentrated in the small area directly around the broken
capsule. The worst that might be expected would be
some damage to soft tissues on the inside of the mouth
and oesophagus as it travelled down to the stomach.

Relatively little of the cyanide salt would have been
transformed to hydrogen cyanide in the mouth; most of
the conversion would have taken place in the very acidic
environment of the stomach before being absorbed into
the body proper. This means that in the mouth and
throat, it would mostly be potassium or sodium cyanide
present, and these salts form an *alkali* solution. This alkali
solution can also burn, as shown by an incident in India.

In 2006, M. P. Prasad added a pinch of potassium
cyanide to a drink, intent on killing himself, but it may
have happened quicker than he planned. He stirred his
poisoned drink with the end of his pen and it is thought
he may have then put the pen to his mouth. As the
poison took effect he had time to scrawl a note: 'Doctors,
potassium cyanide. I have tasted it. It burns the tongue
and tastes acrid.' This is the burning Silva referred to, but

from the potassium cyanide, not the hydrogen cyanide he spoke of. The chemical corrosion of potassium cyanide would also only affect the soft tissues, leaving the bone and teeth intact. It also begs the question that, if the cyanide was concentrated enough to cause a burning sensation and dissolve a big chunk of Silva's bone (although I don't know how), why wasn't it enough to kill him?

Cyanide is such a chemically useful unit that it is found all over the place, including in the food we eat, not just the bitter almonds mentioned earlier that are a particularly rich source of cyanide.[14] Our bodies produce an enzyme that processes this natural cyanide by adding a sulphur atom and converting it into the significantly less toxic thiocyanate. In this way the average human can process up to a gram of cyanide every 24 hours. Problems arise when a large amount of cyanide is ingested in one go and the system is overwhelmed. Perhaps Silva is genetically predisposed to have much larger amounts of the enzyme that does the thiocyanate conversion. Eating a sulphur-rich diet might also help, but there are other ways to minimise the effects of cyanide.

It is theoretically possible that Silva could have inadvertently reduced the effects of the cyanide by eating a lot of sugar. Experiments conducted on rats showed those fed on cyanide and sugar fared a lot better than those fed cyanide alone. It is believed that the cyanide

---

[14] Sweet almonds are fine, in case you were worried. But apricot stones also contain a lot of cyanide and should be eaten only in moderation.

binds to the sugar so it is excreted before it can block the cytochrome oxidase enzyme needed for respiration. It's an interesting theory but it hasn't been proved, and sugar certainly isn't a recognised or recommended antidote for cyanide poisoning.[15]

There are things Silva's captors could have done to prevent him dying from cyanide poisoning, whether it was feeding him a lot of garlic, full of delicious sulphur compounds, or lots of sugar or, more credibly, giving him a cyanide antidote when needed. The effect cyanide is shown to have on Silva's face, I can't explain. And though it is scientifically unrealistic, it certainly makes for a dramatic scene. For a brief moment, the roles are reversed. M is seen as the callous, uncaring one while the main villain is the figure of sympathy, a victim of someone else's plans and ambitions.

---

[15] Please *don't* try this at home.

# 024

# *Spectre* and Bond's backstory

James Bond is an instantly recognisable figure, no matter who is wearing the tuxedo. We feel that we know him, and to a certain extent we do, though his background and inner thoughts are scarcely mentioned. We think we understand his motivations and relationships with those around him. While he's hurled into any number of unusual or unpredictable situations, we have a good idea of how he will respond. But how much do we actually know about him? His well-known trademarks – the vodka Martinis, the gun, the car – are all superficial adornments. Is there any substance to pad out that well-tailored dinner jacket?

Bond emerges from every mission shaken but rarely stirred. Decades of near-death experiences, killing in cold blood and watching friends, colleagues and loved ones die have barely made a dent on his mental health. Several PTSD-worthy events are included in every adventure, but the biggest psychological trauma must be in *Spectre*. How does someone recover from the revelation that all the misery and trauma they have been through has been instigated by one person, and that one person is your adoptive brother who you thought was dead? It would take years of therapy to unpack all that.

We know Bond's body must be built differently to survive the physical traumas he is routinely put through (see chapter 021), but his psychological wiring must also be incredibly robust to withstand the unique pressures of his lifestyle. How does Bond sleep at night let alone buck himself up and stride unflinchingly into the next mission?

## Solid foundations

According to his creator Ian Fleming, Commander James Bond, CMG, RNVR looks like Hoagy Carmichael,[1] but 'with something cold and ruthless in his eyes.' He is 6ft tall with blue eyes, black hair that falls in a comma on his forehead, and a 3in (8cm) vertical scar on his cheek and another scar on his hand, covered up with a skin graft. He was born to Scottish and Swiss parents,[2] both killed in a climbing accident when James was 11. He was raised by an aunt in Kent and spent time in Austria, where he was taught to ski by Hannes Oberhauser, 'a wonderful man' who became 'something of a father' to him. Recruited to the secret service from the navy, he killed a Japanese cipher expert in New York and a Norwegian double agent in Stockholm to earn his double-0 status.

His job requires that Bond's unique set of skills be put into practice on two or three missions a year, followed by a couple of weeks' holiday and any necessary sick leave, which is often a lot. The Bond of the books is dug out of the ruins and carried away in an ambulance

---

[1] An American singer, songwriter and actor, popular through the 1930s to 50s.

[2] He is described as an Englishman in *From Russia with Love*, but the character has been portrayed by American, South African, Irish, Australian and Welsh, as well as English, actors. Fleming gave him Scottish heritage in later novels, after Sean Connery was cast in the role.

(*Moonraker*).[3] His lengthy recoveries, or deprogramming sessions after being brainwashed by the Russians (*The Man with the Golden Gun*), still leave plenty of time for his 'dog days', periods of ennui when he is stuck in the office doing endless paperwork. The novels give him time to reflect on his career as a government assassin pitted against some of the world's greatest criminal masterminds and he acknowledges it doesn't offer much in the way of long-term prospects. He doesn't expect to reach the standard double-0 retirement age of 45 and spends what money he earns and wins in casinos on enjoying himself in the meantime.

His connoisseur's taste in fine food, wines, cars, cigarettes, women and gambling is mostly Fleming's own. The author also donated his public school background, rank as commander, and love of adventure, danger and golf to the fictional character. He donated other less savoury habits too – snobbishness, racism and a dismissive attitude towards women. Fleming took all of this and transported Bond into a life he would like to have lived, full of exotic locations, rich food, beautiful women and thrilling adventures.

The producer Cubby Broccoli observed Fleming would 'have given anything, I imagine, to be James Bond'. But a colleague from Fleming's war years noted, 'He was a pen-pusher like the rest of us. All through the war I thought of him as a collector of rare books. Of course, he knew everything that was going on, but he

---

[3] In the film version of *Moonraker*, Bond is again taken away in an ambulance but only because it is being used by Drax's heavies to kidnap Bond, not because Bond is in need of serious medical attention.

never seemed to show any real inclination to take part in it. If he was secretly longing for action I never saw any sign of it.'

Nevertheless, Fleming was obviously the basic template for his fictional superspy. He also met plenty of other people who provided inspiration and anecdotes that would flesh out the bones and add credibility to such an incredible character. Many of those Fleming met during the Second World War have been suggested as a 'real-life James Bond'.

## Proto-Bond

Dusan 'Dusko' Popov was a charming, wealthy Yugoslav who risked his life to work as a double agent for the British. Like every other German spy in Britain during the war, he was controlled by the XX committee (standing for double cross), part of MI5.[4] Codenamed 'Tricycle', he supplied information about the German secret service to the British and returned false information deliberately designed to confuse and hide the reality of British military operations and capabilities. He insisted his cover would be that of a playboy. He stayed at the best hotels, spent extravagantly and attended the best parties. He liked to say his primary interests were 'sports cars and sporting women'. Though he was often described as 'the real James Bond', he said 007 wouldn't have lasted two days in the real world of espionage.

Though Popov seems an excellent prototype Bond, Fleming didn't know him personally. The spy claimed

---

[4] He was awarded an OBE in 1947 for his services, appropriately enough in the cocktail bar at the Ritz Hotel in Piccadilly.

Fleming had seen him humiliate a wealthy Lithuanian at a gambling table in Lisbon using his knowledge of the casino and a large amount of bravura, a story not a million miles from the premise of *Casino Royale*.[5] However, Fleming had much closer friends and acquaintances that were of a similar mould to Popov.

Commander Wilfred 'Biffy' Dunderdale, the SIS station chief at Paris who drove around in a bulletproof Rolls-Royce, was a great friend of Fleming and claimed to find parts of his own stories in the 007 novels. Commander Alexander 'Sandy' Glenn, an Old Fettesian (like Bond), was an assistant naval attaché in Belgrade who had an affair with the wife of a Belgian diplomat. Gus March-Philips, codenamed W.01, was given command of his own team in West Africa, hence the 'W.', and some have suggested the 01 part of his codename signified a licence to kill.

Perhaps one of the more significant influences was Sir William 'Little Bill' Stephenson, a Canadian in charge of British security coordination for the entire western hemisphere. He was based in New York for a time, where Fleming visited him as part of his liaison work for NID.[6] One 3.00 a.m. during his stay Stephenson, with two assistants and Fleming tagging along, broke into the office of the Japanese consul-general, cracked the safe and copied the codebooks it contained before returning

---

[5] Fleming's account of his time at the Lisbon casino is rather different (see chapter 021).

[6] This is also where Fleming met Colonel William 'Big Bill' Donovan, chief of the US intelligence section. Donovan gave Fleming a .38 revolver engraved with 'for special services' in gratitude for his work.

everything. The story, much elaborated, became James Bond's first assassination on his way to earning his double-0 status.

Stephenson also set up and ran a spy training school at Oshawa, not far from Toronto. It was staffed with experts in firearms, unarmed combat, safe-blowing, lock-picking, explosives, ciphers and everything else a spy might need to be knowledgeable about. Fleming apparently spent a few days there and, if the stories are to be believed, took part in some of the training exercises.[7] Apparently his highest score was obtained in an underwater exercise to attach a limpet mine to a derelict tanker moored in a lake – a similar scenario appears in Fleming's book *Live and Let Die.* Another elaborate test for the recruits was to plant an imaginary bomb at a strategic point in Toronto.[8] Fleming and a few others were assigned the Toronto power station as their target. While the other trainees devised ingenious ways to smuggle themselves inside in coal trucks etc., Fleming put on his best suit and poshest English accent, rang up the managing director, explained he was a visiting British engineer, and arranged an appointment. It's a ploy not dissimilar to the one Goldfinger would use to infiltrate the waterworks near Fort Knox.

Sensational and far-fetched as these events may seem, they did happen, even if Fleming's involvement has been greatly exaggerated. And the people doing them may have

---

[7] Fleming may have visited the facility but there is no evidence that he completed any courses, at least none that the historian David Stafford could find.

[8] The city police and other officials had been warned so as not to cause a major incident.

started off as enthusiastic amateurs in the spirit of fictional spies like Richard Hannay and Bull-Dog Drummond, but .they received training and carried out these missions as their job. The Second World War in particular established the world of espionage on a professional basis. James Bond is part of that professional establishment, even though he frequently bends or breaks the rules. Being 007 is his job, and one that he is very good at.

## Job title

Just as there are many inspirations for James Bond's character, not only those mentioned above, there are almost as many theories as to the origins of his 007 job title. Fleming said in a *Playboy* interview that the idea of 007 came from his time in NID, when all top-secret documents had a double-0 prefix written on the front. The 00 code, signifying 'for your eyes only', has been attributed to John Dee, advisor to Queen Elizabeth I and alleged spy.[9] Another theory is that 007 was based on Fleming knowing that 0070 was the German diplomatic code used to send the Zimmerman (secret diplomatic) telegrams from Berlin to Washington. My personal favourite is my mum's assurance that 007 was the number of the bus that ran between Canterbury and the Kent coast, where Fleming had a house.[10]

---

[9] Dee steeped himself in all kinds of magical and occult theories and apparently considered seven to be a lucky number.

[10] It's a lovely theory but the idea of Fleming getting on a bus, or even paying any attention to public transport when there were fast cars to be driven at dangerous speeds through the Kent countryside, is unbelievable. If you want to catch the 007 bus today it runs between Deal and London Victoria.

The double-0 code may have real-life origins, but the job and licence it gives to James Bond do not. During the Second World War, ideas were put forward to establish a list of assassination targets and to train specialists to kill them. The idea didn't progress into reality for several reasons. It was not out of squeamishness, or the British sense of 'fair play', but because killing people has repercussions. For example, in 1942 SOE-trained Czechoslovakians assassinated Reinhard Heydrich, Protector of Bohemia and Moravia. The Germans thought they had found the assassins and razed their home villages, shot all men over 16 and deported most of the women and children to concentration camps. The other consideration was that assassinations, however well planned, do not always work. Stewart Menzies, head of SIS, noted, 'if assassination were easy many statesmen and high officers would have come to a violent end.' Anyone employed in this kind of work would need to have a unique set of highly developed skills. Someone like James Bond, in fact.

## A stressful job

All of us sometimes feel the pressure of work; however, few of us are trying to meet deadlines and reach attainment targets while being shot at or an industrial laser advances towards our genitals. Yet Bond can face down these terrifying situations with nothing more than a raised eyebrow and a quip? Some people simply don't experience fear in the way others do. Maybe Bond is one of those people.

The amygdala is a region of the brain that processes emotion. Some people who have a damaged amygdala, through injury or surgery, don't feel fear. One particularly well-studied individual, known only as SM, displays a complete lack of fear and has great difficulty in detecting possible threats.[11] And, despite going through a high number of near-death experiences, she is completely immune to post-traumatic stress disorder (PTSD).

PTSD is characterised by reliving a traumatic event from the recent or distant past as flashbacks, nightmares or intense emotional responses to reminders of the event. There can also be depression and increased anxiety, as well as behavioural changes and greater susceptibility to infection. The types of events most likely to cause PTSD are severe threats to life or personal safety, or intentional human-inflicted harm, such as assault or torture, or killing people – basically Bond's job description. Those who have a family history of traumatic events, like losing both parents in a tragic accident at a young age, are at an even greater risk.

In *No Time To Die*, Bond appears to be enjoying a quiet retirement free from PTSD. Scenes of him waking in the middle of the night screaming may have been cut, or Bond has somehow survived his career as a double-0 agent psychologically undamaged. Studies of war veterans have discovered 10 key aspects that have helped cushion them against the worst mentally: training, a

---

[11] In one experiment she visited a 'haunted house' experience with a group of scientists. She voluntarily led the group around, confidently walking down dark hallways and round blind corners, encouraging everyone to follow her. Actors dressed as monsters hid in the house and repeatedly tried to scare her, but she just smiled, laughed and tried to engage them in conversation.

moral compass, a role model, a mission, optimism, head-on confrontation of fear, altruism, humour, social supports and spirituality. James Bond ticks nearly all 10 boxes.

We've already discussed Bond's training in chapter 021. His unwavering moral compass is on display in every film. We know Bond will do anything for Queen, country and all that is right and good, even when he goes rogue and defies M's very specific orders. M, as well as providing a clear mission in every film, may also act as a role model, though Bond is not in such awe of his boss that he doesn't question his instructions, and openly mock him on occasion. Regardless, Bond always enters his missions wholeheartedly and doesn't shy away from any difficulties or dangers, so he certainly has 'optimism' and 'head-on confrontation of fear' covered. His altruism cannot be doubted either, as evidenced by the numerous women he has had to pluck from the dangerous situations he dragged them into, even when it threatens his own personal safety and the outcome of the mission. His quips and jokes are more than just entertainment for the audience, as we have seen in chapter 021, and will also help him mentally. Positive emotions and humour can promote flexible thinking and problem solving, essential when trying to figure out how to escape a criminal mastermind's clutches. By contrast, his spirituality or personal beliefs are never mentioned. There is also a notable absence of social supports in 007's life.[12]

---

[12] I don't think the string of Bond girls he has slept with count in this context, or the occasional drink with Felix Leiter.

Bond works alone. He occasionally teams up with Leiter or Mathis, and sometimes Q and Moneypenny have gone above and beyond to help Bond outside of his official remit, but these people rarely do more than supply him with equipment, information, a new suit or a cover story. The marines or SAS may arrive to help in the final showdown, but they are all nameless faces in a large, well-armed crowd. There is no Watson to Bond's Holmes, no Robin to his Batman, or Penfold to his Danger Mouse. There is no constant companion to help and support him.

Perhaps aware that Bond might appreciate a friend or two – especially when he is up against 'the author of all your pain', Blofeld in *Spectre* – Bond, M, Moneypenny, Q and Tanner team up together.[13] Though they all clearly get on and respect each other, they don't appear to socialise outside of work and Bond doesn't keep in touch with any of them after his retirement. Bond has never been portrayed as very sociable. His abrasive manner may be a deliberate ploy to get under his enemies' skin, but he can't always switch it off when talking to his friends and colleagues. Maybe it's them who don't want to spend time with a trained killer with attachment issues and a high alcohol dependence.

Fleming gave the 007 film producers Saltzman and Broccoli a memorandum that became the definitive thesis on all things Bond, including how he should be played: 'James Bond is a blunt instrument wielded by a

---

[13] Somehow this feels wrong, like the Scooby gang about to pull Blofeld's mask from his face to reveal it had been the janitor all along.

Government Department. He is quiet, hard, ruthless, sardonic, fatalistic ... Neither Bond nor his Chief, M, should initially endear themselves to the audience. They are tough, uncompromising men.' And the filmmakers succeeded. As Nomi says to Moneypenny, 'I can see why you shot him.' Moneypenny's response is equally telling: 'everyone tries sooner or later.' Few people might want to be Bond's friend.[14] And yet there are plenty of people who want to be him or bed him.

## A class act

Much of the attractiveness of James Bond must come down to the lifestyle, locations and look. Whatever mess Bond finds himself in, it is often in a very nice part of the world, and he always carries out his missions with effortless cool, charisma and style. When Sean Connery was cast in the role he was a relatively unknown actor, but his looks and physicality won over the producers. Connery, immaculately turned out in Savile Row suits, gave an assured performance in *Dr No* that won over everyone else, including Fleming, who was initially unimpressed with the Scottish actor.[15]

George Lazenby, a former model, in his only outing as 007 shows a far more vulnerable Bond, one who has reached the end of his considerable talents and must be rescued by Tracy di Vicenzo. Was this one of the factors that made him less popular in the role?[16]

---

[14] Bond, for his part, seems to be perfectly happy without them.
[15] Allegedly his exact words were 'that fucking truck driver'.
[16] Or was it the ruffled dress shirt and kilt combo?

Roger Moore, like Connery, is never ruffled. He is wittier but not as physically menacing as Connery. Had Connery walked into the Harlem branch of the Fillet of Soul in *Live and Let Die* we would be expecting a fight. Moore, in his immaculate coat and shiny shoes, just looks lost. That is not to say Moore's Bond did not have the ruthless edge Fleming considered so important. Both Connery and Moore stretched the bounds of what was considered 'fair play'. In *Dr No*, Connery shoots an unarmed man in the back. In *For Your Eyes Only*, Moore is faced with a henchman trapped in a car teetering on the edge of a cliff and gives it a fierce kick to send him to his inevitable death.

Perhaps in response to Moore's perceived lighter touch, Timothy Dalton's Bond was billed as a grittier incarnation of 007.[17] His portrayal was closer to the original Bond of Fleming's stories. He hardly smiles, remains very focused and has notably less success with women than his cinematic predecessors, though this is not necessarily down to character development alone. When *The Living Daylights* was released in 1987, the threat of AIDS was particularly prominent, and casual sex, even in the fictional world of 007, was a greater taboo than it had been. Despite Bond's reputation, Dalton's amorous restraint is actually closer to the character of Fleming's novels, who averaged one conquest per adventure.

By the time Pierce Brosnan put on the black tie and fired his PPK down the gun barrel, the world had

---

[17] Dalton's films still have plenty of fun moments: the exploding milk bottles, the revolving sofa in Q's workshop, winning the cuddly toy at the funfair.

changed again. Six years had passed since Dalton had thrown in his gambling chips, the Berlin Wall had fallen and there was a woman in charge of MI6. Bond was starting to look like, as M put it, 'a sexist, misogynist dinosaur, a relic of the Cold War'. He still uses his 'boyish charms' to get what he wants, though it doesn't always work, either on M or the other women he would encounter. Not much else changed though. Brosnan's Bond is still invincible, still well tailored and still has time for a quip, a drink and a visit to the casino.

Until Daniel Craig was cast in the role, the films didn't bother with Bond's backstory or any degree of soul-searching, with the possible exception of 1969's *On Her Majesty's Secret Service*. Rather than isolated missions, the Craig-era films show some of the impact of such a stressful way of life and hint at explanations for this serious, slightly detached character.[18] There has at least been an acknowledgement that a person can't just walk away from Bond's experiences unscathed. The death of Vesper Lynd in *Casino Royale* makes Bond cold and vengeful in *Quantum of Solace*. In *Skyfall* he revisits childhood traumas at his family home and his much-loved boss dies in his arms. The big revelation in the following film, *Spectre*, is that everything Bond has gone through in the past three films has been orchestrated by his arch-nemesis, and it all stems from a severe case of sibling rivalry. Bond and Blofeld are stepbrothers.

After being orphaned, Bond was adopted by Hannes Oberhauser and gained a brother, Franz Oberhauser.

---

[18] Though one man's cool and aloof is another man's antisocial arsehole.

Forced to spend time with this overachieving outsider, Franz used his jealous hatred as fuel for his plans for world domination, terrorism, revenge, extortion, and so on. Franz kills his father, fakes his own death and reinvents himself as the big, bad Bond villain Ernst Stavro Blofeld. Bond doesn't take it too well either, and tries to kill his stepbrother on several occasions. Despite this animosity, the pair are always remarkably polite to one another in conversation. Just because you are a criminal mastermind or professional assassin, doesn't mean you can't have manners.

Whatever Bond is, he is not neurotypical. How that affects him, who can say – the filmmakers haven't let us in to see his inner thoughts. It has certainly been an advantage to MI6 and M, who have exploited his mental toughness with extraordinary success.

# 025

# *No Time to Die* and the nanobots

Most Bond villains are usually only in it for themselves. Lyutsifer Safin, by contrast, is just trying to help, even if no one wanted or asked for that help. Sure, he wants to kill people – lots of people – but if we only listened to his calm reasoning we would realise that dying horribly was actually in our best interests. Safin has decided that humanity is stuck in a rut and that killing millions of people will provide an evolutionary kick up the backside that will bring about a new world order. There are no ransom demands, no highly profitable contracts to take out specific individuals, just death, and as much of it as possible. Safin's lack of personal gain is unusual by Bond villain standards. The high-tech method he has chosen for his diabolical plan is, however, entirely in keeping with the franchise.[1]

The sinister weapons in *No Time to Die* are nanobots, tiny devices that can be 'DNA programmed' to kill only the people you don't like, which in Safin's case seems to be almost everyone. Mass slaughter is nothing new in the 007 franchise. Blofeld, Stromberg and Drax have all been willing to destroy humanity as a whole (see chapter 006), and Safin has taken some inspiration from all of them.[2] The 25th Bond film contains a

---

[1] As are the gaping plot holes.

[2] And from Fleming's novel *You Only Live Twice*, where Blofeld has designed a poison garden to entice people to kill themselves. In the

staggering number of references to previous films, and also breaks a lot of the franchise's unwritten rules. But the long-standing tradition of stretching scientific credibility up to and beyond its limits isn't one of them.

The premise of the film is that Lyutsifer Safin, understandably upset at the poisoning of his entire family by SPECTRE agent Mr White, decides to take revenge. He turns up at Mr White's house but Mr White isn't home, so instead he kills Mrs White but saves young Madeleine White.[3] This simple idea, a loose thread at the start of the film, rapidly unravels into a complicated plot about an even more complicated weapon.[4]

Having killed one person and saved another, Safin appears to do nothing but seethe in the background for the next 20 years.[5] I'm sure he didn't just sit about twiddling his thumbs but, like any good megalomaniac, he was just waiting for the right moment and the right weapon to come along. He discovers the perfect weapon in a top-secret laboratory. Every villain has to have a secret lab, but MI6 has them too.

---

film, Safin has a poison garden inherited from his father and has decided to export death rather than waiting for people to turn up on his doorstep.

[3] Who changes her name to Madeleine Swan and falls in love with James Bond.

[4] I miss the days of rich bad guy doing expensive evil to enrich himself further. Life, and 007 plots, were simpler back then.

[5] In the intervening years, Mr White, the man Safin wanted to kill, is poisoned by his former employees, SPECTRE, and kills himself (*Spectre*, see chapter 004). Maybe that's why Safin is so upset.

## Secret labs and secret weapons

The secret laboratory we see at the start of *No Time to Die* is no ordinary secret laboratory. It is so secret that even MI6 doesn't know about it, just M, the people who work there and a few master criminals such as Blofeld and Safin.[6] The research being undertaken on a high floor of a skyscraper in south London is into biological weapons – Ebola, smallpox, those kinds of deadly things, things so awful there are international treaties banning their use (see chapter 006 for more biological nasties).[7] There is also Project Heracles, the target for the SPECTRE henchmen that come abseiling through the windows.

Heracles, son of Zeus, was the greatest hero of Ancient Greek mythology. He demonstrated incredible feats of strength and bravery, killing evil creatures and saving the innocent from marauding monsters. It's an appropriate name for what M and the weapon's creator, Obruchev, claim is the perfect weapon, able to kill the bad guys but leave the innocent untouched. It's even more appropriate if you read to the end of Heracles's story, where he is accidentally killed by a poisoned shirt that tears the flesh from his bones. But we are getting ahead of ourselves.

Back in the office block in south London, a vial of Heracles, the scientist Obruchev and a USB memory stick full of DNA data are taken from the laboratory and

---

[6] Is MI6 suddenly that bad at finding out secret stuff now that Bond has retired? Even the CIA have an inkling of the lab and its contents.

[7] Why in the name of health and safety have they decided to use an office block in a densely populated city for their investigations into highly contagious, lethal pathogens?

whisked off to Cuba for a party. The party is being held in honour of Blofeld's birthday,[8] and it's where we get to see the first demonstration of the weapon's capabilities. Every SPECTRE agent is there, as are MI6 and CIA agents sent to recover Obruchev and the mysterious weapon. Bond, working for the CIA, is also mingling in the crowd. Obruchev, meanwhile, is busy upstairs programming the vial of the secret weapon, though he surreptitiously switches memory sticks to use a different set of data. After programming, the contents of the vial are launched into the ventilation system to rain down on the partygoers downstairs. All the SPECTRE agents fall to the ground, blood and blisters appearing on their faces, but everyone else is unharmed.

An extended gun–fight and a few explosions later, and Bond has recovered the memory stick but lost Obruchev, who is abducted again but this time by Safin's men.[9] With no one around to explain what the hell is going on, Bond turns to Q for answers.

## Listen to the experts

It takes seconds for Q to figure out it's all about nanobots, or nanorobots, tiny devices between 0.1 and 100 micrometers in size, or about as big as a red blood cell. We see a picture of one on Q's computer screen, a tiny spider or tick-like thing clamped onto a cell of

---

[8] We know the date is 28 May as Blofeld shares his birthday with his creator, Ian Fleming.

[9] Why did Safin need Obruchev abducted from Cuba along with his briefcase? He already had everything set up on his island. If he didn't want MI6 or the CIA getting information out of him, why didn't they just kill him?

some kind (I couldn't tell what) with a red blood cell in the background for scale.[10] Along with the helpful image, the memory stick also has DNA profiles of lots of people.[11] It seems these robots can be programmed to recognise whole DNA profiles or short sequences of DNA to select individuals, or groups of people with the same genetic characteristic, and kill them. This is how the SPECTRE agents were killed at the party but everyone else was unaffected. Q also reveals that Bond, having been exposed to nanobots in Cuba, now has them for life.

In another exposition dump a few scenes later, we learn that these nanobots had been under development for the past 10 years and that it was M who had supported it, unofficially and completely off the books. A weapon that can be programmed to target only certain people is only as good as the person doing the programming. M may have wanted to take out assassins and master criminals but what if it is the master criminals doing the programming? To make matters worse, these microscopic killing machines can also be passed on to anyone by touch. Those who attended the funerals of the SPECTRE agents killed in Cuba also die after touching their bodies. The DNA is similar enough to trigger the nanobots' murderous rampage through their relatives' bodies as well. M and the scientists involved in the project don't seem to have thought this through very well. M's defence is that he had instigated

---

[10] Red blood cells are 6–8 micrometers in size, which is 6,000–8,000 nanometers (nm), or 0.006cm.

[11] Moneypenny chips in that DNA databases around the world have been hacked into, and presumably Obruchev or Safin has been collecting the results.

and supported the research with the best of intentions, to keep the UK safe. So I guess that's OK then.

A lot of information is divulged very quickly and it raises a lot of questions that the film doesn't have much time to answer between the fight, chase and action sequences that follow. Setting aside questions about the ethics, culpability and massive lack of foresight that these nanobots pose, what about the practicalities? How are they made? How are they programmed? How are they powered? How do they infect people? How do they kill? Will you die after shaking hands with someone because you happen to have a similar section of DNA to someone somebody wants dead? In short, should you be worried? The short answer is no. You can rest easy because nanobots aren't actually real, certainly not in the way they are presented in *No Time to Die*, at least not yet.

## There's plenty of room at the bottom

In a lecture delivered in 1959, the physicist Richard Feynman speculated about miniaturisation – not just making things smaller but building things up from the very bottom, literally atom by atom. If components measuring tens or hundreds of atoms across could be made and joined together to build a working car, it would be just 1mm (0.04in) across, maybe even smaller. Today we would call this nanotechnology because the components are measured on the nanoscale,[12] but the word wasn't used until much later.[13]

---

[12] 1 metre is 1,000,000,000 nanometres.

[13] Though well received by the audience, the lecture 'There is Plenty of Room at the Bottom' had little scientific impact at the time. It

Biology is already well ahead of us, as Feynman pointed out in his lecture. A single cell can store all the information needed to create a fully functioning organism and still have room for a power system, manufacturing of components, monitoring systems and so much more. Feynman asked his audience to 'Consider the possibility that we too can make a thing very small which does what we want – that we can manufacture an object that manoeuvres at that level!'

Scaling robots down to a size where they could fit inside a human vein and travel round the body would be a massive leap forward in health care. Drugs could be delivered precisely where they are needed; important processes could be monitored without blood tests or biopsies; cancerous cells could be destroyed without invasive surgery or the need for rounds of chemo and radiotherapy. These and many more possibilities are why nanotechnology is a very active area of scientific interest and not just in medicine.

The miniaturisation of devices means that they can potentially be used unnoticed, making them particularly attractive for espionage work. As well as using micro-scopic robots with cameras or listening equipment, they could also be used as trackers. With Bond being so prone to wandering all over the world and finding himself in all sorts of inhospitable situations, it's useful for his employers to be able to keep tabs on him. Radio transmitters have been hidden in his shoes and he has

---

was rediscovered in the 1990s, when nanotechnology was gaining serious attention and at least some of what the Nobel laureate had speculated about 30 years earlier was proving to be possible.

swallowed radioactive pills (see chapter 002), but in *Casino Royale* the technology has been shrunk to a size that can be injected. It works for pets that get lost or stolen, so why not secret agents?

Well, for one, because you need a scanner to read your pet's chip. The microchip is passive until it is powered by an external source, usually the scanner, that also picks up the signal that is sent back to identify the dog or open the cat flap. M's pet agent is supposedly trackable from anywhere in the world, so there must be some kind of power source incorporated into the chip so it can send a signal for MI6 to pick up. Though why they need to bother when angry calls from foreign ambassadors and the trail of destruction Bond inevitably leaves in his wake should be able to pinpoint his location pretty easily is another matter.

In *Spectre*, the tracking technology has shrunk even further and new features have been added. Bond gets injected with smart blood, which Q says is the latest in nanotechnology and can monitor his vital signs as well as locate him. It worked so well in that film, it is used again in *No Time to Die*, just before Nomi and Bond are sent off to infiltrate Safin's lair.[14]

While scientists have come a long way since Feynman's lecture and can do extraordinary things on a very small scale, they can't yet do all the things needed to produce something as sophisticated as smart blood or Project

---

[14] If Nomi has been a double-0 agent for two years, why is she only getting her injection now, and at the last minute? And why does Q never explain anything until after he has given the injection? Has he not heard of informed consent or do secret agents not enjoy that right? I have so many questions.

Heracles. Q-branch and the scientists working on Project Heracles are way ahead of their time. Nanotechnology presents a lot of exciting possibilities but also many technical challenges.

## Small steps

One big problem is size. Making something on the nanoscale requires specialist techniques. There are two approaches: top-down or bottom-up. Bulk materials can be broken up into smaller pieces, and atoms or molecules can also be reacted or brought together in controlled ways to build up bigger structures. There are a number of different ways of doing either of these things but none of them are easy or quick. These individual components also have to interact with each other if the nanobot is going to work as a whole, so there is little margin for error. And the more complex the tasks you need your nanorobot to carry out, the more components you are going to need. The nanobots in *No Time to Die*, whether they are smart blood nanobots or Project Heracles nanobots, are going to need to be very complex indeed.

In the case of the smart blood, if you want feedback on what is going on inside the body, the nanobots are going to have to have an array of biosensors. Fortunately many already exist, and even in suitably small sizes. For example, gold nanoparticles can be coated with flu antibodies so if the flu virus is present in a sample, it will bind to the nanoparticles, changing their colour from red to blue. Other research has taken individual enzymes involved in glucose conversion and connected them to carbon nanotubes that act as wires. When an enzyme converts a molecule of glucose, a tiny electrical signal travels down the wire. This could form the basis of a device for real-time monitoring of blood glucose levels for diabetics. Many of these devices are still in the experimental stage and none, so far, have been tested inside the human body. Monitoring vital signs is theoretically possible, but there is still the problem of transmitting this information to Q circling thousands of feet up in an aeroplane, as discussed earlier.

The Project Heracles nanobots aren't concerned with vital signs (the very opposite, in fact). Nor do they need to transmit any information to anyone. But they do need to be very manoeuvrable. The human body is significantly compartmentalised and there are barriers and membranes to keep things separate. Smart blood is injected, presumably directly into the bloodstream so these nanobots can travel around the body within the circulatory system. The weaponised nanobots, however, have to get inside the body all by themselves if they are to be passed on by the slightest touch. Getting through the skin is no mean feat because one of the skin's major roles is to keep things out of the body, and it is very good at it. These nanobots will also need to get back out again so they are on the skin's surface, ready to move on to the next person.

As an aside, if nanobots can get in and out of the body so easily, why can't they be removed? Why is Bond, and everyone else who encounters him and anyone else carrying the nanobots, stuck with them for life? If there is a minimum number of nanobots needed to kill, how many times can someone pass them on before they become too diluted to be a problem? We know that at least two transmissions are possible because nanobots that were sprayed onto Madeleine's wrist are passed on to Bond and then on to Blofeld, killing him in seconds.[15]

Perhaps those little leg-like structures can help the nanobots claw their way past the tightly packed cells that form these membranes. And perhaps those same legs might also be used to puncture cells and destroy them. It would mean one nanobot could continue puncturing cells as fast as its little legs will carry it, making them potentially very dangerous even in tiny amounts. It might also explain the bleeding we see on the nanobots' victims when they die.

Moving parts, however, need energy to power them, perhaps from an onboard battery or from metabolising nutrients from their surroundings. Maybe there are cleverly designed molecular levers that are triggered by the interaction with certain chemicals in the body. Whichever way it's done, it must be controllable so the nanobots only puncture the cells in their selected targets.

Programming the nanobots could be done in a number of ways. For instance aptamers, small sections of DNA that act like antibodies binding to specific proteins, can be synthesised in labs. Picking the right proteins could make the weapon as discerning as the film suggests. A selection of major histocompatibility complex proteins

---

[15] I'm so sad that Blofeld, one of the best loved of the Bond villains, dies in such an unimpressive way, quickly and quietly off screen. A character as eccentric and over-the-top as Ernst Stavro Blofeld deserved an extravagant and over-the-top death scene.

(MHC) might do the trick. These are protein markers presented on cell surfaces to identify it as part of the body and not an invader. These are the proteins screened when looking for matches for organ transplants, to prevent the recipient's immune system from attacking the donor organ. Analysing a DNA profile for a unique set of MHC then making the aptamers to match them would be difficult, time-consuming and costly, and still wouldn't work as Safin wants because these aptamers get broken down in the body over time.[16]

On top of all this, all the materials used to make these nanobots have to be non-toxic and biocompatible to avoid poisoning or provoking an immune response in everyone as soon as they enter the body. Using biological materials – such as DNA, for example – as a physical scaffold is possible, but they are likely to be degraded by the body. If Safin's nanobots are for ever, inert materials like gold could be used that will stay in the body for a long time because there is no biological means of breaking them down and excreting them.

Having made all these biocompatible components, there is then the problem of assembling them into a working nanorobot. With a lot of planning and care, it would theoretically be possible to design the components with pre-prepared sections that snap together so they could assemble themselves. This high level of sophistication is currently well beyond today's technological capabilities,

---

[16] Incidentally, how is Safin funding this secret facility manufacturing mass slaughter? His father was a SPECTRE's poisoner. Did he inherit the facility when his parents were killed by Mr White? Were the huge sums needed to convert it into a nanobot factory also inherited? How much does a SPECTRE poisoner earn?

but the nanobots' designers must have found a way because Safin, or his minions, appear to be making them in a big pond with nothing more sophisticated than a few lights and some stirring.

The pond is in the middle of Safin's lair, a former chemical weapons factory, and it looks none too clean in my honest opinion, adding the complication of potential contaminants into the mix. There is a shiny spotless laboratory as well, full of glass tanks, more trays of liquid, lots of lights and only a couple of computers. The minimalist look fits Safin's general aesthetic but the distinct lack of any laboratory equipment suggests the nanobots have been so well designed they can do almost all the work themselves.

Bond villains generally like to make life, and death, complicated. If there is a more convoluted path to their diabolical goal, they will take it. But Lyutsifer Safin is a model of minimalist efficiency. He lets MI6 do all the difficult research and development work, and has Obruchev, his scientist on the inside, feeding him all the information he needs to mass-produce his own nanoweapons. He lets SPECTRE steal the weapon, the scientist and the DNA data, and then uses them to kill his most villainous opposition. All he has to do is step in at the last minute to gather everything together, scale up production and sort out distribution, which amounts to chartering a couple of boats. The nanobots will then spread themselves.

Once Nomi and Bond have broken into Safin's lair, they soon realise the scale of his operation and how easily it could become a global disaster. Killing everyone in sight quickly shuts down production, but there are

still the two boats coming to collect the first shipments
of nanobots. There are also various governments growing
increasingly concerned at the amount of activity
happening on and around a politically disputed island.
Instead of stopping the two boats and quietly dismantling
the facility, M on Bond's suggestion, orders missiles to be
launched from a UK ship to destroy everything in one
big ball of flames. Apparently creating an international
incident in this act of aggression is a better option than
risking anyone else finding out about M's staggeringly
poor judgement.[17]

By blowing everything up, the filmmakers continue
with the tradition of ending things with a bang, but they
also killed Bond in the process. Sixty years of life-
threatening missions, and five bullets fired into him as he
scampered around Safin's lair, couldn't stop him, but the
missile strikes from HMS *Dragon* can. But, for once, he
wasn't even trying to escape. Having been infected with
nanobots programmed with Madeleine Swan's DNA,
just a hug with her or her daughter would kill them
both. In terms of nanobots Bond has quite the collection.
Two injections of smart blood, one set of nanobots from
the party in Cuba, another from Madeleine's perfume
spray, and a third set from Safin's ornamental pond. The
danger he poses probably doesn't stop at Madeleine and
their daughter.

Bond is dead and so is his friend Felix Leiter. So are
his enemies, and his enemy's enemies.[18] But everyone

---

[17] How is this man still in his job?

[18] Are the filmmakers, like Safin, wiping the slate clean for a new
generation of Bond films?

else is safe because, once again, Bond saves the day and the world from another evil plot to destroy humanity. The science underpinning that plot is shaky at best, but does the science have to be factual in an otherwise fictional universe? I suppose it depends on how realistic the story needs to be. However, it doesn't even obey its own internal logic. But logic, or lack of it, is yet another Bond tradition.

In a film that contains so many of the well-worn tropes that we have come to love and expect in a 007 adventure, there are some major breaks with tradition. Unlike any other James Bond story, he makes the ultimate sacrifice. Yet Bond can never really die. Like Dr Who or the British monarchy, he will regenerate in a new form. Instead of a proclamation from the palace gates, 'Bond is dead. Long live James Bond', the well-known phrase appears at the very end of the credits: 'Bond will return.'

# Bibliography

## Published works

Akhavan, J. 2015. *The Chemistry of Explosives* (3rd edn). RSC Publishing, Cambridge.

Amis, K. 1968. *Colonel Sun*. Random House, London, 2015.

Antman, E. M., Smith, T. W. 1985. Digitalis Toxicity. *Annu Rev. Med.*, 36: 357–367.

Ashcroft, F. 2012. *The Spark of Life: Electricity and the Human Body*. Penguin Books, London.

Babu, R. 2006. The Only Taste: Cyanide is Acrid. *Hindustani Times*, 8 July, https://www.hindustantimes.com/india/the -only-taste-cyanide-is-acrid/story-vhsbYsiNyWzIfak N4HBK0H.html

Benson, R. 1984. *The James Bond Bedside Companion: The Complete Guide to the World of 007*. Boxtree Limited, London.

Bereanu, V., Todorov, K. 1994. *The Umbrella Murder*. Pendragon Press, Cambridge.

Berkhouse, L., Davis, S. E., Gladeck, F. R., Hallowell, J. H., Jones, C. B., Martin, E. J., Miller, R. A., McMullan, F. W., Osborne, M. J. 1983. *Operation Dominic I – 1962*. Defense Nuclear Agency, Washington, DC.

Berntson, G. G., Bechara, A., Damasio, H., Tranel, D., Cacioppo, D. C. 2007. Amygdala Contribution to Selective Dimensions of Emotion. *SCAN*, 2: 123–129.

Black, J. 2001. *The Politics of James Bond: From Fleming's Novels to the Big Screen*. Praeger Publishers, Westport, CT.

——. 2021. *The World of James Bond: The Lives and Times of 007*. Rowman and Littlefield Publishing Group Inc., Lanham, MD.

Bradley, D. 2017. VX Nerve Agent Behind Kim Jong-nam's Murder. *Chemistry World*, 24 February, https://www .chemistryworld.com/news/vx-nerve-agent-behind-kim -jong-nams-murder/2500460.article

Brake, M. 2020. *The Science of James Bond: The Super-Villains, Tech, and Spy-Craft Behind the Film and Fiction*. Simon & Schuster, New York.

Bray, C. 2010. *Sean Connery: The Measure of a Man*. Faber & Faber Ltd., London.

Broccoli, C., Zec, D. 1998. *When the Snow Melts: The Autobiography of Cubby Broccoli*. Macmillan Publishers Ltd., London.

Bucaretchi, F., de Deus Reinaldo, C. R., Hyslop, S., Madureira, P. R., De Capitani, E. M., Vieira, R. J. 2000. A Clinico-Epidemilogical Study of Bites by Spiders of the Genus *Phoeutria. Rev. Inst. Med. Trop. S. Paulo*, 42(1): 17–21.

Burns, J. F. 1982. 2 Scandals Have All Moscow Abuzz. *The New York Times*, 27 February, Section 1, Page 3.

Bush, S. P., King, B. O., Norris, R. L., Stockwell, S. A. 2001. Centipede Envenomation. *Wilderness and Environmental Medicine*, 12: 93–99.

Caldicott, D. G. E., Croser, D., Manolis, C., Webb, G., Britton, A. 2005. Crocodile Attack in Australia: An Analysis of Its Incidence and a Review of the Pathology and Management of Crocodilian Attacks in General. *Wilderness and Environmental Medicine*, 16: 143–159.

Canale, L., Comtet, J., Niguès, A., Cohen, C., Clanet, C., Siria, A., Bocquet, L. 2019. Nanorheology of Interfacial Water During Ice Gliding. *Physical Review X*, 9, 041025.

Cassidy, C., Doherty, P. 2017. *And Then You're Dead: A Scientific Exploration of the World's Most Interesting Ways to Die*. Allen & Unwin, London.

Chapman, J. 1999. *Licence to Thrill: A Cultural History of the James Bond Films*. I. B. Tauris Publishers, New York.

Chippaux, J.-P., Goyffon, M. 2008. Epidemiology of Scorpionism: A Global Appraisal. *Acta Tropica*, 107: 71–79.

Clegg, B. 2012. *Armageddon Science: The Science of Mass Destruction*. St. Martin's Press, New York.

Cook, T. 2015. Ground-Shaking Research: How Humans Trigger Earthquakes. *Earth Magazine*, 3 April, https://www.earthmagazine.org/article/ground-shaking-research-how-humans-trigger-earthquakes

Cork, J., Stutz, C. 2015. *James Bond Encyclopedia*. Penguin Random House, London.

Coupe, J., Bardeen, C. G., Robock, A., Toon, O. B. 2019. Nuclear Winter Responses to Nuclear War Between the

United States and Russia in the Whole Atmosphere
Community Climate Model Version 4 and the Goddard
Institute for Space Studies ModelE. *Journal of Geographical
Research: Atmospheres*, 8522–8543.

Creer, B.Y., Smedal, H. A., Wingrove, R. C. 1960. *Technical Note
D-337: Centrifuge Study of Pilot Tolerance to Acceleration and
the Effects of Acceleration Pilot Performance*. National
Aeronautics and Space Administration, Washington, DC.

Dash, A., Mohapatra, P. C. 2013. *Nano Robotics – A Review*.
Conference paper, National Conference in Modern Trends
in Engineering Solutions.

De Santana, C. D., Crampton, W. G. R., Dillman, C. B., et al.
2019. Unexpected Species Diversity in Electric Eels with a
Description of the Strongest Living Biolelectricity
Generator. *Nature Communications*, 10: 4000.

Diels, J.-C., Arissian, L. 2011. *Lasers: The Power and Precision of
Light*. Wiley-VCH Verlag GmbH & Co., Weinheim.

Ding, J., Chua, P.-J., Bay, B.-H., Gopalakrishnakone, P. 2014.
Scorpion Venoms as a Potential Source of Novel Cancer
Therapeutic Compounds. *Experimental Biology and Medicine*,
0: 1–7.

Egan, Sean. 2006. *James Bond: The Secret History*. Kings Road
Publishing, London.

Emsley, J. 1998. *Molecules at an Exhibition: The Science of Everyday
Life*. Oxford University Press, Oxford.

Erickson, G. M., Gignac, P. M., Steppan, S. J., Lappin, A. K.,
Vliet, K. A., Brueggen, J. D., Inouye, B. D., Kledzik, D.,
Webb, G. J. W. 2012. Insights into the Ecology and
Evolutionary Success of Crocodilians Revealed Through
Bite-Force and Tooth-Pressure Experimentation. *PLosONE*,
7(3): e31781.

Feinstein, J. S., Adolphs, R., Damasio, A., Tranel, D. 2011. The
Human Amygdala and the Induction and Experience of
Fear. *Current Biology*, 21: 34–38.

Feynman, R. 1960. There's Plenty of Room at the Bottom.
*Engineering and Science*, 23: 22–36.

Field, M., Chowdhury, A. *Some Kind of Hero: The Remarkable
Story of the James Bond Films*. The History Press, London.

Fleming, F. (Ed). 2015. *The Man with the Golden Typewriter: Ian Fleming's James Bond Letters*. Bloomsbury Publishing, London.

Fleming, I. 1953. *Casino Royale*. Penguin Books Ltd., London, 2006.

———. 1954. *Live and Let Die*. Penguin Books Ltd., London, 2006.

———. 1955. *Moonraker*. Penguin Books Ltd., London, 2006.

———. 1956. *Diamonds Are Forever*. Penguin Books Ltd., London, 2006.

———. 1957. *From Russia, with Love*. Penguin Books Ltd., London, 2006.

———. 1958. *The Diamond Smugglers*. Vintage Books, London, 2009.

———. 1958. *Dr No*. Penguin Books Ltd., London, 2006.

———. 1959. *Goldfinger*. Penguin Books Ltd., London, 2006.

———. 1961. *Thunderball*. Penguin Books Ltd., London, 2006.

——— 1962. *For Your Eyes Only*. Penguin Books Ltd., London, 2006.

———. 1962. *The Spy Who Loved Me*. Penguin Books Ltd., London, 2006.

———. 1962. *Thrilling Cities*. Vintage, London, 2013.

———. 1963. *On Her Majesty's Secret Service*. Penguin Books Ltd., London, 2006.

———. 1964. *You Only Live Twice*. Penguin Books Ltd., London, 2006.

———. 1965. *The Man with the Golden Gun*. Penguin Books Ltd., London, 2006.

———. 1966. *Octopussy*. Penguin Books Ltd., London, 2006.

Fruck, L., Kerbs, A. 2021. *Bionanotechnology: Concepts and Applications*. Cambridge University Press, Cambridge.

Fry, B. G., Wroe, S., Teeuwisse, W., van Osch, M. J. P., Moreno, K. *et al.* 2009. A Central Role for Venom in Predation by *Varanus Komodoensis* (Komodo Dragon) and the Extinct Giant *Varanus* (*Megalania*) *Priscus*. *PNAS*, 106(22): 8969–8974.

Funnell, L. (Ed.). 2015. *For His Eyes Only: The Women of James Bond*. Wallflower Press, London and New York.

Gauglitz, G. G., Korting, H. C., Pavicic, T., Ruzicka, T., Jeschke, M. G. 2009. Hypertrophic Scarring and Keloids: Pathomechanisms and Current and Emerging Treatment Strategies. *Mol. Med.*, 17(102): 113–125.

Greenemeier, L. 2007. TASER Seeks to Zap Safety Concerns. *Scientific American*, 3 December, https://www .scientificamerican.com/article/taser-electric-shock-zap -law-canada/

Gresh, L. H., Weinberg, R. 2006. *The Science of James Bond: From Bullets to Bowler Hats to Boat Jumps, the Real Technology Behind 007's Fabulous Films*. John Wiley & Sons, Inc., Hoboken, NJ.

Grubich, J. R., Huskey, S., Crofts, S., Orti, G., Porto, J. 2012. Mega-Bites: Extreme Jaw Forces of Living and Extinct Piranhas (Serrasalmidae). *Scientific Reports*, DOI: 10.1038/srep01009.

Harding, L. 2016. *A Very Expensive Poison: The Definitive Story of the Murder of Litvinenko and Russia's War with the West*. Guardian Books, London.

Harkup, K. 2015. *A is for Arsenic: The Poisons of Agatha Christie*. Bloomsbury Publishing, London.

Harris, R., Paxman, J. 2011. *A Higher Form of Killing: The Secret History of Chemical and Biological Warfare*. Random House, London.

Hart, M. 2001. *Diamond: The History of a Cold-Blooded Love Affair*. Fourth Estate, London.

Heard, B. J. 2008. *Firearms and Ballistics: Examining and Interpreting Forensic Evidence*. John Wiley & Sons Ltd., Chichester.

Hess, W. N. 1964. *The Effects of High Altitude Explosions*. NASA Technical Note D-2402. National Aeronautics and Space Administration, Washington, DC.

Hillman, H. 1993. The Possible Pain Experienced During Execution by Different Methods. *Perception*, 22: 745–753.

Holden, A. 1995. *The St Albans Poisoner*. Corgi Books, London.

Inglis, L. 2018. *Milk of Paradise: A History of Opium*. Pan Macmillan Ltd., London.

Jeffery, K. 2011. *MI6: The History of the Secret Intelligence Service 1909–1949*. Bloomsbury Publishing, London.

Jenner, R., Undheim, E. 2017. *Venom: The Secrets of Nature's Deadliest Weapon*. Natural History Museum, London.

Kemp, A. 2013. Diamonds Are a Laser Scientist's New Best Friend. Phys. Org., 7 August, https://phys.org/news/2013 -08-diamonds-laser-scientist-friend.html

Kerr, M., Rodriguez McRobbie, L. 2021. *Ouch! Why Pain Hurts and Why it Doesn't Have to.* Bloomsbury Sigma, London.

Lindner, C. (Ed.). 2003. *The James Bond Phenomenon: A Critical Reader.* Manchester University Press, Manchester.

Lycett, A. 2012. *Ian Fleming: The Man Who Created James Bond.* Hachette UK, London.

McKeever, W. 2020. *Emperors of the Deep: The Mysterious and Misunderstood World of the Shark.* HarperCollins Publishers, London.

McNess, A. 2015. *A Close Look at 'A View to a Kill'.* Printed by Amazon.

Merkle, S. W. 1997. Non-Nuclear EMP: Automating the Military May Prove a Real Threat. *Military Intelligence Professional Bulletin,* https://irp.fas.org/agency/army/mipb/1997-1/merkle.htm

Miller, R. 2010. *Codename Tricycle: The True Story of the Second World War's Most Extraordinary Double Agent.* Random House, London.

Miodownik, M. 2013. *Stuff Matters: The Strange Stories of the Marvellous Materials that Shape Our Man-Made World.* Penguin UK, London.

Moore, R. 1973. *The 007 Diaries: Filming Live and Let Die.* The History Press, London, 2018.

———. 2008 *My Word is My Bond: The Autobiography.* Michael O'Mara Books Limited, London.

Muolo, M. J. 1993. *Space Handbook: A Warfighter's Guide to Space, Volume 1.* Air University Press, Maxwell Air Force Base, AL.

Osterweis, M., Kleinman, A., Mechanic, D. (Ed.). 1987. *Pain and Disability: Clinical, Behavioural, and Public Policy Perspectives.* National Academy Press, Washington, DC.

Pearson, J. 1966. *The Life of Ian Fleming.* Bloomsbury Publishing Ltd., London, 2013.

———. 1976. *James Bond: The Authorised Biography.* The Random House Group Limited, London, 2006.

Peterson, M. E. 2006. Black Widow Spider Envenomation. *Clinical Techniques in Small Animal Practice,* 21(4): 187–192.

Pieretti, S., Di Giannuario, A., Di Giovannandrea, R., Marzoli, F.,
    Piccaro, G., Minosi, P., Aloisi, A. M. 2016. Gender Differences
    in Pain and Its Relief. *Ann Ist Super Sanità*, 52(2): 184–189.
Prahlow, J. 2010. *Forensic Pathology for Police, Death Investigators,
    Attorneys, and Forensic Scientists*. Springer, London.
Queiroz, H., Magurran, A. E. 2005. Safety in Numbers?
    Shoaling Behaviour of the Amazonian Red-Bellied Piranha.
    *Biology Letters*, 1: 155–157.
Reuben, C. 2012. The Atlas-1 Trestle at Kirtland Air Force Base,
    NM, https://ece-research.unm.edu/summa/notes/trestle.html
Robinson, J. 2005. Rumble in the Jungle With Amazon's Killer
    Piranha. *Los Angeles Times*, 22 November, https://web
    .archive.org/web/20110831110223/http://travel.latimes
    .com/articles/la-os-piranha22nov22
Saukko, P., Knight, B. 2004. *Knight's Forensic Pathology* (3rd edn).
    Hodder Arnold Ltd., London.
Sellers, R. 2019. *When Harry Met Cubby: The Story of the James
    Bond Producers*. The History Press, Cheltenham.
Smith, E. A. 2020. *Ian Fleming's Inspirations: The Truth Behind the
    Books*. Pen and Sword History Books Ltd., Philadelphia, PA.
Smith, G. 2021. *Overloaded: How Every Aspect of Your Life is Influenced
    by your Brain Chemicals*. Bloomsbury Sigma, London.
Snyder, R. G. 1963. *Human Survivability of Extreme Impacts in
    Free-Fall*. Federal Aviation Agency, Aviation Medical Service,
    Aeromedical Research Division, Civil Aeromedical
    Research Institute, Oklahoma City, OK.
Soares, S., Sousa, J., Pais, A., Vitorino, C. 2018. Nanomedicine:
    Principles, Properties, and Regulatory Issues. *Frontiers in
    Chemistry*, 6: 360–375.
Sompayrac, L. 2008. *How the Immune System Works* (3rd edn).
    Blackwell Publishing, Malden, MA.
Stone, T. and Darlington, G. 2000. *Pills, Potions and Poisons*.
    Oxford University Press, Oxford.
Streatfeild, D. 2002. *Cocaine*. Virgin Books Ltd., London.
Swee, G., Schirmer, A. 2015. On the Importance of Being Vocal:
    Saying 'Ow' Improves Pain Tolerance. *The Journal of Pain*,
    16(4): 326–334.

Tilney, N. L. 2003. *Transplant: From Myth to Reality.* Yale University Press, New Haven, CT and London.

Times Wire Services. 1989. Jet Liner Rips Open; 11 Die. *Los Angeles Times,* 24 February, https://www.latimes.com/archives/la-xpm-1989-02-24-mn-420-story.html

Tolan, M., Stolze, J. 2020. *Shaken, Not Stirred!: James Bond in the Spotlight of Physics.* Springer Nature, Cham.

Varnau, G. J., de Wit, H. 2020. Diamond Raman Lasers Offer Multifaceted Potential. *Photonics Spectra,* https://www.photonics.com/Articles/Diamond_Raman_Lasers_Offer_Multifaceted_Potential/a65991

Vittitoe, C. N. 1989. Did High-Altitude EMP Cause the Hawaiian Streetlight Incident? *System Design and Assessment Notes,* Note 31.

Waring, R. H., Steventon, G. B., Mitchell, S. C. (Ed.). 2002. *Molecules of Death.* Imperial College Press, London.

Weiss, D., Tomasallo, C. D., Meiman, J. G., Alarcon, W., Graber, N. M., Bisgard, K. M., Anderson, H. A. 2017. Elevated Blood Lead Levels Associated with Retained Bullet Fragments – 2003–2012. *Morbidity and Mortality Weekly Report,* 66(5): 130–133.

Winder, S. 2007. *The Man Who Saved Britain: A Personal Journey into the Disturbing World of James Bond.* Pan Macmillan Ltd., London.

Wroe, S., Huber, D. R., Lowry, M., McHenry, C., Moreno, K., Clausen, P., Ferrara, T. L., Cunningham, E., Dean, M. N., Summers, A. P. 2008. Three-Dimensional Computer Analysis of White Shark Jaw Mechanics: How Hard Can a Great White Bite? *Journal of Zoology,* 1–7.

Yeung, A. S., Ivkovic, A., Fricchione, G. L. 2016. *The Science of Stress: What it is, Why We Feel it, How it Affects Us.* Ivy Press, Brighton.

Zhu, Z., Song, W., Burugapalli, K., Moussy, F., Li, Y.-L., Zhong, X.-H. 2010. Nano-Yarn Carbon Nanotube Fibre Based Enzymatic Glucose Biosensor. *Nanotechnology,* 21: 165501.

# Acknowledgements

James Bond doesn't always save the world single handed. He has a team of people helping in the background, supplying him with all the necessary tools and logistics support needed to thwart criminal masterminds. Writing a book isn't all down to one person either. Thanks must go first of all to my bosses, or Ms, Jim Martin and Angelique Neumann at Bloomsbury.

Many others have acted like Felix Leiter. Chris Culnane, David and Sharon Harkup, Matthew May and Mark Whiting have all helped tidy up the mess I've created and smooth over my more disastrous writing. David Jesson, Claire Benson, Laurie Winkless and Michael Williams have stood in for Q with technical support. Justin Browers has provided a sympathetic ear and supplied me with papers with all the consideration and professionalism of Miss Moneypenny.

And, like the elite fighting force that gives Bond tactical support in the big showdown against the baddie, there are The Marmalades (Jodie Eastwood, Julia Graves, Odette Brady, Gary Glass and Atika Shubert), with brilliant feedback, encouragement and support throughout the whole writing process. A special mention has to go to Bill Backhouse who has put up with an awful lot of Bond-related questions and listened patiently to my random Bond-related theories. As always, my parents, Margaret and Mick, have uncomplainingly (at least to me) read every word I have written and given helpful feedback. A big thank you to all of them. Any remaining mistakes, broken bits or plot holes are all down to me.

# Index

acetylcholine, 55–56
actors, 359–62
Adam, Ken, 70, 74, 178–80
aerial spins, 150–52
aircraft, getting sucked out
     of, 291–97
alcohol, 313
alternating current (AC), 173
Amasova, Anya, 209–10
ammunition, 134–39
amygdala, 309, 355–56
animals, 111
   bioelectric capabilities
      of, 174–78
   blue-ringed octopus, 32
   categories, 112–13
   stereotyping, 115–19
anthrax, 88
Archimedes, 41
Armendáriz, Pedro, death
     of, 191–92
Aston Martin DB5
   gadgets of, 216–19
   lasers used by, 49–50
Atlantis, 74–77
atomic explosions, 158, 186–87
   detonating devices, 195–96
   disassembling bombs, 196–97
   electronic side effects, 256–59
   gold radiation, 191–94
   plutonium-239, 187–89
   radiation damage, 189–91
   realising, 187–89
   Red Snow, 194–96
   reshaping plutonium, 199–200

atoms, breaking/making bonds
     between, 158–62

*Bhagavad Gita*, 187
binary weapon, 57–58. *See also*
     VX, never agent
bioterrorism, 83–84
   background on, 84–85
   modifying plot
      involving, 88–92
   omega virus, 88–92
   orchids, 92–96
   ricin, 94–96
   villain plan involving, 85–88
black piranha *(Serrasalmus
     rhombeus)*, 120
black powder, 159
Black, Honor, 206–7
Blaize, John, 106
Blofeld, Ernst
   bioterrorism plot
      of, 83–96
   constructing volcano lair
      for, 70–74
   scars of, 278
body temperature, 324–25
Bond, James
   actors playing, 359–62
   backstory of, 348–62
   basic template for, 349–51
   car protection, 141
   companionship for, 358–59
   exepreincing stress, 355–59
   and *Goldfinger* laser, 43–47
   gun-barrel sequence, 9–23

inspiration for job title
    of, 354–55
medical treatment for, 314–16
mental health of, 309–11
physical fitness of, 306–8
physical/mental health
    of, 304–5
portraying inner thoughts
    of, 359–62
proto-Bond, 351–54
wildlife connection to, 113
Boothroyd, Geoffrey, 129,
    138–39
Bottome, Phyllis, 12
Bouvier, Pam, 140
Bray, Hilary, 16
British Intelligence, 12
Broccoli, Cubby, 70
Brosnan, Pierce, 37, 155, 167,
    199, 211–13, 305, 342,
    360–61
bulletproof shield, Aston Martin
    DB5, 217–18
bullets, 135–36. See also
    ammunition; guns
butterfly act, 58–60

carbon
    atoms, 98
    dioxide, 45, 98, 158–59
    fibre, 264
cardiac arrhythmias, 174
cars. See Aston Martin DB5;
    vehicles
    protection from bullets, 141
    remote-controlled, 271–73
Carver, Elliott, 261, 261–67. See
    also stealth
Case, Tiffany, 102
Casino Royale, 304–5

knotted rope in, 316–19
poisoning drinks in, 60–64
TV special, 17
writing, 13–17
castor oil, 94–96
Castro, Fidel, 65
cello case, chase sequence
    involving, 221–26
centipedes, 114
centrifuge, 306–7
Chandler, Henry, 202
charm offensive, 209–11
chase sequences, 215–16
    cello case sequence, 219–26
    foot chases, 226–29
    in Goldfinger, 216–19
    tank, 219–20
chemical explosions, 158–62
chemicals, 92–96
Churchill, Robert, 29, 129
Citroën 2CVs, 225
coca plants, 242–43
cocaine, 240–45
coding machine, 34–35.
    See also From Russia with
    Love
Cold War, 19–20
Collard, John, 106–7
Colt Model 1908 Vest
    Pocket, 133–34
compressed-gas bullets, 300–301
Connery, Sean, 30, 35–36, 46,
    114, 122, 203, 206, 270,
    305, 339, 359–60
Conqueror, The, 191
coronavirus, 19
cortisol, 310
Cotton, Sydney, 29
Craig, Daniel, 154, 305, 361–62
Crandall Canyon Mine, 252–53

CRISPR (clustered regular
    interspaced short palindrome
    repeats), 283–84
crocodiles, 124–26
Cuneo, Ernie, 16
cyanide, 335–36
    antidotes to, 338–39
    science of, 336–39
    various form of, 339–47

D'Abo, Maryam, 223
Dahl, Roald, 207–8
Dalton, Timothy, 49–50, 175,
    223, 240, 360
death, methods of, 320–21
    in aquarium, 299–300
    body covered in crude
        oil, 321–22
    boundary between atmosphere
        and space, 297–301
    compressed-gas bullets,
        300–301
    crushing enemies with
        thighs, 301–3
    cyanide pill, 335–47
    disrupting oxygen
        cycle, 333–34
    fires, 328–31
    getting sucked out of
        aircraft, 291–97
    gold-paint death, 322–28
    liquid nitrogen, 331–33
debris, space, 79–80
defibrillator, 181–82
Dennis-Forbes, Ernan, 12
diacetylmorphine, 236
diamonds, 97
    carbon atoms in, 98
    ending literary association
        with, 106–7

forming of, 98–99
as high-status objects, 99–100
as personal
    adornments, 107–10
in private homes, 100
prototype laser, 103–5
synthetic, 104–5
temptation to steal, 100–101
Diamonds are Forever, 51, 97
    adapting plot of, 101–5
    car chase in, 220
    inspiration for diamonds
        in, 97–101
Die Another Day, 51
    cyanide in, 342–43
    death of villain in, 295–97
    defibrillator in, 181–82
    diamonds in, 109–10
    lack of parachutes in, 150
    robot suit in, 182–85
digitalis poisoning, 60–64
direct current (DC), 173
dirty bomb, 192–93
Disco Volante, 165–66
disguises, 261–67
dopamine, 232
007. See Bond, James
double-0 code, 354–55
Dr No, 187
    animals featured in, 113–15
    cyanide used in, 339–40
    dragon tank in, 261–62
    establishing rules, 17–22
    explosion in, 162–64
    fire-related death in, 328–31
    gun-barrel sequence
        in, 9–10
    guns in, 130–31
    portrayal of women
        in, 203–6

radiation damage in,   189–91
setting stage for,   12–17
teaser line,   22–23
toxins in,   33–34
Drax, Hugo,   47, 83, 111
   biology-inspired plan
      of,   92–96
   death of,   297–99
   defining features of,   276
   exploding space station
      of,   157–70
   lair of,   77–82
   stealth of,   262
drinks, poisoning,   60–64
drugs,   230–31
   heroin,   235–36
   in *Live and Let Die*,   236–40
   morphine,   234–36
   opium,   233–36
   reward pathway,   231–33
dry skin,   176–77
Dunderdale, Wilfred 'Biffy',   352
Dutch gold,   326


E
earthquakes, creating,   251–54
Eichengrün, Arthur,   235
ejector seat,   218
electric chair, execution
      by,   178–80
electric eels,   174–75
electricity
   and EM pulse bomb,   254–
      59
   principles of,   172–74
electrocution, death by,   171–72
   in *Die Another Day*,   180–85
   electric chair,   178
   electricity principles,   172–74
   water,   174–78

electromagnetic pulses (EMP).
      *See* EM pulse bomb
EM pulse bomb,   246–47
   science behind,   254–59
   sound as weapon,   249–51
Escobar, Pablo,   244
explosions,   157–58
   basic principles of,   158–62
   device components,   162
   flammable materials
      in,   162–67
   getting creative with,   167–68
   in space,   169–70


F
Faraday, Michael,   184
Feynman, Richard,   368
fibrillation,   62–63
Fiedler, Richard,   328
fight-or-flight response,   309–11
firearms. *See* guns
fires,   228–31
flammable materials,   162–67
flammables, hoarding,   164–65
Fleming, Ian
   animals inspiring,   113–15
   appropriation by,   15–17
   diamonds inspiring,   97–101
   drinking habit of,   314
   early life of,   11–12
   idea for golden girl,   322–28
   interest in guns,   128–31,
      138–39
   interest in unusual
      weapons,   247–49
   literary association with
      diamonds,   106–7
   postwar writing,   13–15
   relationships with
      women,   202–3

researching
  bioterrorism, 85–88
  in Second World War, 12–13
  as template for Bond, 349–51
  topicality in books of, 26–28
  winter sports, 224
foot and mouth disease, 89–90
*For Your Eyes Only*, 111
  chase sequence in, 225
  death by electrocution
    in, 171–85
  headphone electrocution
    in, 180–81
Frazer-Smith, Charles, 29,
  101–2
*From Russia with Love*, 24–26
  blade-in-shoe science, 31–36
  hoarding flammables
    in, 164–65
  preparing right equipment
    for, 26–28
  Q-Branch in, 28–31
  toys for boys, 37–39
  women in, 205–6

gadgets
  preparing right
    equipment, 26–28
  Q inspiration, 28–31
  in *From Russia with
    Love*, 24–39
  toys for boys, 37–39
Gale, Cathy, 206–7
gamma gas, 52–58
  poisoning drinks, 60–64
  and ricin, 58–60
  thallium poisoning, 64–67
gene therapy, 281–84
glass, 141
glasses, X-ray, 37–38

Global Positioning System
  (GPS), 36
gold paint, death by. *See*
  Masterson, Jill
*GoldenEye*, 39, 69
  EM pulse in, 246–59
  grand plan in, 247
  plane stunt in, 149
*Goldfinger*, 36, 40–41, 187
  Aston Martin DB5
    in, 216–19
  breakdown of laser
    in, 43–47
  death of villain in, 293–95
  death rays, 41–43
  electric heater in, 175–76
  electrocution in, 177–78
  flexible explosive in, 161
  gold-paint death in, 322–28
  guns in, 133
  rendering gold worthless
    in, 191–94
  women in, 203, 206–7
Goldfinger, Auric, 40
Goldfinger, Erno, 16
gondola/hovercraft, 268–69
Goodhead, Holly, 47, 157–58,
  210
Graves, Gustav, 41, 150
  and gene therapy, 281–84
  robot suit of, 182–85
gravity, 79, 147, 223
grays (Gy), 190
grotto sharks. *See Thunderball*:
  sharks in
gun barrel, film sequence, 9–11
  establishing rules
    for, 17–22
  setting stage for, 12–17
  teaser line, 22–23

guns
    ammunition used for,   134–37
    crafting,   128–31
    principles of,   131–34
    protection using,   139–42
    silencers,   137–39

haemolacria,   280–81
Hall, Tracy,   104
Hauschild, Fritz,   312
headphones, electrocution
        by,   180–81
Hearst, William Randolph,   263
*Her Majesty's Secret Service*,   208
Hercules Motorized Traction
        Table,   248–49
heroin
    converting morphine
        into,   238–40
    origins of,   235–36
Heydrich, Reinhard,   355
highs, crushing enemies
        with,   301–3
Hillman, Harold,   179
histocompatibility complex
        proteins (MHC),   373–74
Hoffman, Felix,   235
Hofmann, Richard,   285
Human Genome Project,   281–82
hydrofoils,   267–69
hydrogen,   159–60, 166
hydrogen cyanide,   336, 341,
        344–46

improvisation, weapons,   246–47
    earthquakes,   251–54
    EM pulse bomb,   254–59
    inspiration,   247–49
    sound as weapon,   249–51
industrial laser. *See* lasers

International Space Station
        (ISS),   78, 169–70, 297
iodine,   193

James Bond, film series
    *Casino Royale*,   304–19
    *Diamonds are Forever*,   97–110
    *Die Another Day*,   291–303
    *Dr No*,   9–23
    establishing rules for,   17–22
    *GoldenEye*,   246–59
    *Goldfinger*,   40–51
    *On Her Majesty's Secret
        Service*,   83–96
    *Licence to Kill*,   230–45
    *Live and Let Die*,   111–26
    *The Living Daylights*,   215–29
    *The Man with the Golden
        Gun*,   127–42
    *No Time To Die*,   363–77
    *Quantum of Solace*,   320–34
    *From Russia with Love*,   24–39
    setting scene for,   12–17
    *Skyfall*,   335–47
    *Spectre*,   348–62
    *The Spy Who Loved
        Me*,   143–56
    teaser line,   22–23
    *Thunderball*,   52–67
    *Tomorrow Never Dies*,   260–73
    *A View to a Kill*,   201–14
    *The World is Not
        Enough*,   274–90
    *You Only Live Twice*,   68–82
    *For Your Eyes Only*,   171–85
Jaws, henchman,   123
jetpack, personal,   269–71
Jinx,   180–85
Jones, Grace,   155
Jong-nam, Kim,   54–55

Kananga, doctor
    animals of,   124–26
    death of,   300–301
    drugs and,   236–40
Karpe, Eugene,   27
Kevlar,   139–40. See also guns:
        protection using
Khan, Kamal,   197–98
Kieselguhr,   160
King, Robert,   167–68
Klebb, Rosa,   206
    inspiration for,   27–28
    shoe used by,   31–36
Kleinflammenwerfer,   328–29
knotted rope,   316–19
Kolevatov, Anatoly A.,   108
Kutsova, Anna,   26

lairs,   68–69
    in space,   77–82
    undersea lair,   74–77
    in volcano,   70–74
Largo, Emilio,   53, 267–69
lasers,   40–41
    acronym,   43
    color of,   46
    death rays and,   41–43
    and diamonds,   101–5
    film usage of,   43–47
    as franchise symbol,   51
    in *The Living Daylights*,   49–50
    in *Moonraker*,   47–49
    photons in,   42–43
    various sizes of,   49–50
lava,   73–74
Lazenby, George,   154, 359
Le Chiffre
    haemolacria of,   280–81
    and knotted rope,   316–18
    poisoning drinks,   60–61

lead,   134–35
Lehder, Carlos,   244
Leiter, Felix,   241
*Licence to Kill*,   111
    aquarium death in,   299–300
    cocaine in,   240–45
    eels in,   174–75
    gun protection in,   140
    huge volumes of fuel in,   166
    sharks in,   122–23
*Liparus*,   74–75
liquid nitrogen,   331–33
Litvinenko, Alexander,
        poisoning,   63–64
*Live and Let Die*,   37, 69, 111–13
    boat stunt in,   150
    compressed-gas bullets
        in,   300
    crocodiles in,   124–26
    sharks in,   124
    sound as weapon in,   249–51
*Living Daylights, The*
    cello case chase sequence
        in,   221–26
    lack of parachutes in,   150
*Living Daylights, The*,   49–50,
        215–16
Llewelyn, Desmond,   31
Lucas, Frank,   237

M,   15, 132, 181, 195, 202, 204,
        208, 229, 305, 317, 342–47,
        357–61, 365
magnesium,   134
Maiman, Harold,   43
*Man with the Golden Gun,
        The*,   127–28
    guns in,   133–34
*Man with the Golden Gun, The*
    aerial spin stunt,   151–52

Mansfield, Cumming,　28–29
Markov, Georgi, death
　　of,　59–60
Masterson, Jill,　320–21
　gold-paint death of,　322–28
　setting stage for death
　　of,　321–22
Maxwell, Robert,　266
May Day,　201, 210–11
Medding, Derek,　169
Menzies, Stewart,　15
Merkle, Scott W.,　257–58
milk of paradise. *See* poppies,
　　extracting opium from
Milovy, Kara,　221–22
miniaturisation,　39, 368–71
Mollaka,　226–29
Moneypenny,　203–5, 266, 335,
　　358–59
Monsieur Diamont,　107
*Moonraker*,　111
　bioterrorism plot in,　92–96
　cyanide used in,　342
　explosions in,　157–70
　lasers in,　47–49
　parachuting in,　146–48
　stealth in,　262
　villain lair in,　77–92
　women in,　210
Moore, Roger,　152, 155, 209,
　　268, 341, 360
morphine,　234–36
Moscow State Circus,　107–8
nanobots,　363–64
　biocompatible components
　　of,　374–75
　miniaturisation and,　368–71
　programming,　373–74
　removing,　372–73
　smart blood,　370–72

specialist techniques for
　　making,　371–77

Naval Intelligence Division
　　(NID),　12
nicotinic receptors,　313
Nieman, Albert,　242
nitrogen,　158–60
nitrogen gas (N2),　159
*No Time To Die*,　363–64
　nanobots in,　366–68
　psychological damage
　　portrayed in,　356–57
　secret laboratory in,　365–66
　tomb explosion in,　162
　women in,　214
Nobel, Immanuel,　160
Noriega, Manuel,　19
nuclear reactors,　163–64

O'Toole, Plenty,　102
*Octopussy*,　198–99
　atomic explosions
　　in,　186–200
　diamonds in,　107
　disarming bombs in,　197
　political context,　197
　stunts in,　152–54
Oddjob, henchman,　177–78,
　　218, 249, 284
oil/squirter,　217
omega virus,　88–92
*On Her Majesty's Secret
　　Service*,　16, 24
　bioterrorism in,　83–96
　chemical plans after,　92–96
one-man helicopter,　271
Operation Starfish
　　Prime,　256–59
opiates,　234–35

opium, 233–36
Oppenheimer, Robert, 187
*Orchidae nigra*, 92–96
organophosphorus
    poisoning, 55–58
oxygen, 33, 36, 45, 155,
    158–60, 169–70, 217, 270,
    292, 294, 299, 300, 306,
    322, 330, 333–34

pain
    dialing down perception
        of, 311–14
    reacting to, 287–90
parachute jump, 143–44
    accidents, 155–56
    avoiding actor deaths, 154–56
    editing stunts, 152–54
    general lack of
        parachutes, 150
    mock-up sets, 155
    in *Spy Who Loved Me* opening
        scene, 144–49
Pauley, Paul, 76
Perry Submarines and
        Oceanographic, 269
photons, 42–43
physical explosions, 158–62
piranhas, 120–21
pistols, 132–33. *See also* guns
plutonium-239, 187–89
poisons, 32–34
Popov, Dusan 'Dusko,' 351–52
poppies, extracting opium
        from, 233–36
post-traumatic stress disorder
        (PTSD)., 356
Prasad, M. P., 345–46
Project Heracles, 365–66, 372
Project Znamya, 42

propellant, 137. See also
        ammunition; bullets; guns
psuedomamma, 280
Pussy Galore, 131, 207

Q-branch, 28–31, 37–38,
        47–50, 92, 216, 218, 225,
        273, 284, 342, 371
Q, inspiration for, 28–31
*Quantum of Solace*
    crude-oil death in, 321–22
    flammable hotel in, 166–67
    parachuting in, 148–49
Quarrel, henchman, 239

radio jamming, 80–82
re-breather, 35. See also
        *Thunderball*
respiration, 337
reward pathway, 231–33,
        312–13
ricin, 94–96
Rigg, Diana, 208–9
Rigoni, Paulo, 225
Rocket Belt, 270
Ryder, Honey, 203–6

Safin, Lyutsifer, 363–64,
        375–77
Sakata, Harold, 284–85
Sanchez, Franz, 165–66,
        241–45
Saunders, Joan, 86
Scaramanga, Francisco, 127
    bullet used by, 136–37
    third nipple of, 280
    weapon used by, 133–34
scars, 276–79
*Scolopendra*, 114
*Sea Shadow*, 261–67

Second World War,  12–13,
     19–20, 93–94, 354
  bioterrorism in,  87–88
  developing explosive
     compounds in,  161
  and double-0 code,  354–55
  L-pills,  338
  radio jamming in,  80–82
Secret Intelligence Services
     (SIS),  12, 352, 355
sharks,  121–24
shock,  173–74
shoe knife, preparing right
     equipment,  24–28
Sierra Leone, diamonds
     in,  108
silencers,  137–39
Sillitoe, Percy,  102
Silva, Raoul,  336, 342–46
silver, reactivity of,  326–27
skiing,  224–25
skin, pores of,  324
Skyfall,  36, 69
  cyanide in,  342–47
  explosions in,  167
smart blood,  370–72, 376
Smith & Wesson Airweight,  129
smoke bombs,  217
snow,  222–23
sound, weaponizing,  249–51
space
  dying in,  291–97
  mirror,  41–42
  villain lair in,  77–82
Spang, Seraffimo,  101
Special Air Service (SAS),  308
Special Operations Executive
     (SOE),  12–13
Spectre,  39, 168, 216, 304,
     348–49

thallium in,  66
  tracking technology in,  370
SPECTRE,  53, 171, 230, 365–66
spiders,  114–15, 203
Sputnik 1,  80
Spy Who Loved Me,
     The,  143–44, 187
  cyanide in,  341–42
  disassembling nuclear missile
     in,  167
  opening scene of,  144–49
  sharks in,  123
  villain lair in,  74–77
  women in,  209–10
staff, recruiting,  73–74
stealth,  260–61
  disguises,  261–67
  heat,  265
  reflecting radio waves,  264–65
  remote-control car,  271–73
  stealth boat logistics,  261–67
  style vs. substance,  267–69
  traveling light,  269–73
stealth boat,  261–67. See also
     Tomorrow Never Dies
Stephenson, William 'Little Bill,'
     352–54
steroids,  285–87
Stolze, Joachim,  147
stress, experiencing,  355–59
Stromberg, Karl
  and atomic bombs,  196–97,
     341
  undersea lair of,  74–77
  webbed fingers of,  279–80
Sub-Biosphere 2, 76–77
sweat,  176–77
Sylvester, Rick,  144–45
syndactyly,  279–80
syringe, hypodermic,  235

tank, escaping in,   219–20
tasers,   183
Tee-Hee, henchman,   125–26
Tellistock Mountain,   149
tetrodotoxin,   32–34
Textron, Bell,   270
.38 Colt Police Positive,   129
.357 Smith & Wesson
      Magnum,   129
Thunderball,   52, 187
   cyanide used in,   340
   execution by electric chair
      in,   178–80
   exploding Disco
      Volante,   165–66
   gamma gas in,   52–58
   origins of,   17–18
   personal jetpack
      in,   269–71
   Red Snow,   194–96
   sharks in,   121–22
   style vs substance
      in,   267–69
   women in,   207–8
tobogganing,   224–25
Tolan, Metin,   147
Tomorrow Never Dies
   stealth boat in,   260–73
   women in,   213–12
topicality,   19–20
tracer, radioactive,   36. See
      also Thunderball
trestle,   256
Trevelyan, Alec
   plan of,   254–59
   scars of,   278–79
trinitrotoluene (TNT),   161

uranium,   162–63
urea,   167–68

vehicles. See cars
   blowing up,   165–66
   disguising,   261–67
   film stunts,   149–52
   jetpack,   269–71
   stealth boat,   260–73
   tank,   219–20
vest, bulletproof. See bullets; guns:
      protection using
View to a Kill, A,   111, 155
   parachuting in,   148
   portrayal of women
      in,   201–14
   ricin used in,   58–69
   triggering earthquake
      in,   251–54
   women in,   210–11
villains,   274–75
   Dr No,   18–22
   enormous strength
      of,   284–87
   gene therapy,   281–84
   pain signals to,   287–90
   physical features,   279081
   scars given to,   276–79
   ways of killing,   291–303
Vincenzo, Tracy di,   208–9
Virulent Newscastle Disease
      (VND),   89
viruses,   88–92
volcano, building lair in,   68–69
   logistics,   70–73
   recruiting staff,   73–74
Volpe, Fiona,   207
VX, nerve agent,   54–58

Wai Lin,   213–14, 266
Walther PPK,   130
Walther WA 2000 rifle,   131–32
water

building lair in,  74–77
electrocution and,  174–78
Wilson, Michael G.,  253
winter spots,  224
*Wizard of Oz, The*,  326–27
women,  201–2
  charming,  209–11
  creating better roles
    for,  211–14
  crushing enemies with
    thighs,  301–3
  in first James Bond
    films,  202–7
  learning from previous
    portrayals of,  207–9
*World Is Not Enough, The*,  187

explosion in,  167–68
reshaping plutonium
  in,  199–200

Yip, David,  155
*You Only Live Twice*,  68–69, 154
  aerial cameramen in,  156
  one-man helicopter in,  271
  piranhas in,  120–21
Young, Arthur Edward,  100
Young, Terence,  207

Zao, henchman,  110, 183
Zokas, Victor 'Renard,'  275
Zorin, Max,  251–54
  steroids used on,  285–87